dinosaurs

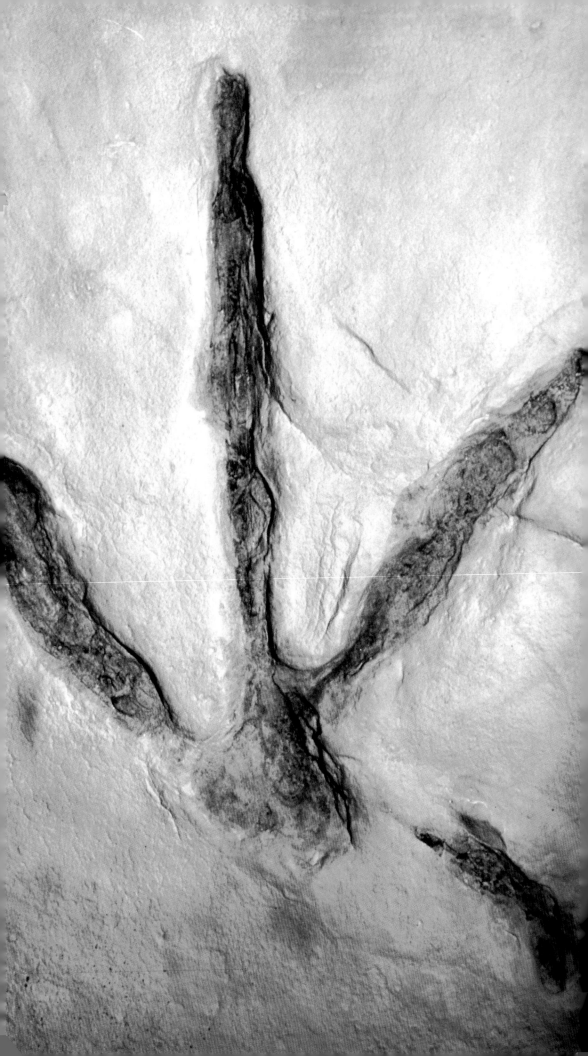

dinosaurs

CHRISTOPHER A. BROCHU, JOHN LONG, COLIN MCHENRY,
JOHN D. SCANLON, PAUL WILLIS

CONSULTANT EDITOR
MICHAEL K. BRETT-SURMAN

FOG CITY PRESS

Published by Fog City Press
814 Montgomery Street
San Francisco, CA 94133 USA

Copyright 2000 © US Weldon Owen Inc.
Copyright 2000 © Weldon Owen Pty Limited
Revised edition 2007

GROUP CHIEF EXECUTIVE OFFICER John Owen
CHIEF EXECUTIVE OFFICER Terry Newell
PUBLISHER Sheena Coupe
CREATIVE DIRECTOR Sue Burk
PRODUCTION DIRECTOR Chris Hemesath
SALES MANAGER Emily Jahn
VICE PRESIDENT INTERNATIONAL SALES Stuart Laurence
ADMINISTRATOR INTERNATIONAL SALES Kristine Ravn

PROJECT EDITOR Robert Coupe
DESIGNERS Lena Lowe, Hilda Mendham
JACKET DESIGN Kelly Booth
PICTURE RESEARCH Annette Crueger
CONSULTANT, REVISED EDITION Dr John Long

ISBN:
Flexibound edition 978-1-74089-584-2
Paperback edition 978-1-74089-583-5
Hardcover edition 978-1-74089-577-4

Printed by Kyodo Printing Co. (S'pore) Pte Ltd
Printed in Singapore

A Weldon Owen Production

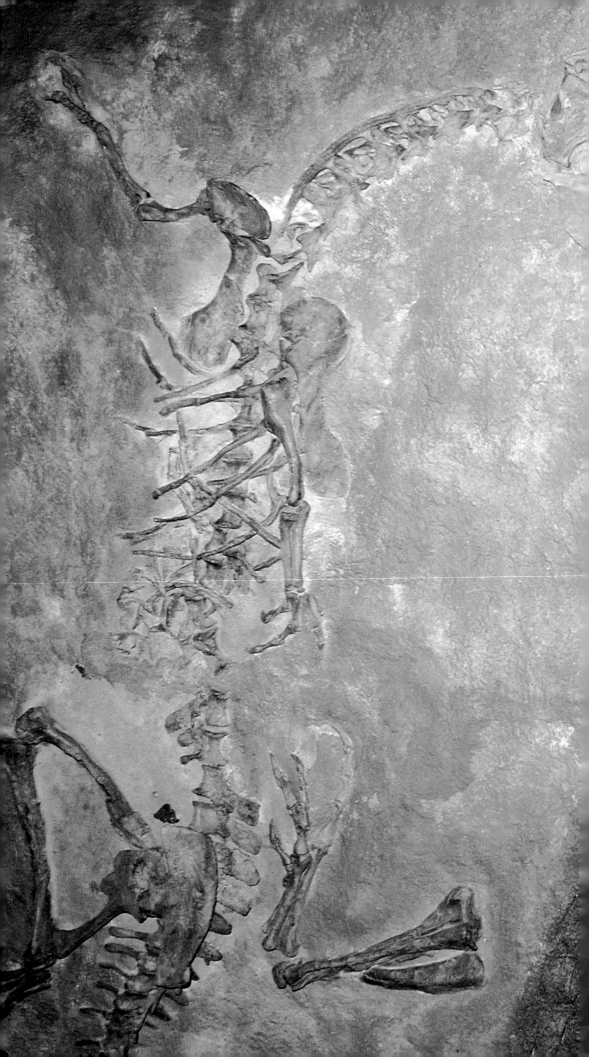

When people call this beast to mind

They marvel more and more.

The Bad Child's Book of Beasts,
HILAIRE BELLOC (1870–1953),
French-born English writer

C O N T E N T S

FOREWORD

When Richard Owen coined the term "Dinosauria"
(meaning "fearfully great, a lizard"), he deliberately
emphasized the superlative form of the ancient Greek word
Deinos, as Homer also did in the *Iliad*. His intent was to focus
on the awe-inspiring aspect of dinosaurs. Owen could not
have known that this group of unusual reptiles would become
part of the world's cultures, the focal point of inspiration for
countless children in every country, and the cure for "science
phobia" in students and adults. Biology and geology, when
combined into the science of paleontology ("discourse on
ancient things"), suddenly becomes a time-travel adventure,
rather than a list of countless unpronounceable Latin and
Greek names. Many discover that learning is fun, and that it
is possible to work names such as Gondwana into conversation
without feeling embarrassed. The key to understanding why
dinosaurs are so popular is that dinosaurs are real, and so was
the lost world they lived in. In one group we can experience
detective and forensic science, raw power, time travel,
adventure, and discovery.

With tongue-in-cheek thanks to Hollywood, professional
dinosaur paleontologists have inadvertently become the
dinosaur's high priests. It is up to us to separate dogma and
just-so stories, and let the truth rise to the surface. Education
for its own sake is not enough. The public must know not
only what happened in the Mesozoic era, but how, why,
and the evidence behind our conclusions.

Paleontology is full of missteps, blind alleys, and discoveries
yet to be made. This book highlights the worldwide search by
many scientists and amateurs for information about the most
absorbing animals in Earth history. Come join our search . . .

M. K. BRETT-SURMAN PH.D.
Consultant Editor

DELVING INTO THE DINOSAUR PAST

Over almost two centuries the science of paleontology has painted for us an increasingly clear and diverse picture of the world that dinosaurs lived in.

To many people, the very word "dinosaurs" conjures frightening images of colossal predatory beasts roaming ancient landscapes, the undisputed masters of the Earth. This popular—and, in the case of most dinosaurs, very inaccurate—perception no doubt helps to explain the peculiar fascination that these long-vanished creatures hold for us. There are, however, many other groups of animals that we know, from the fossil record, to have been contemporary with the dinosaurs, and many of these were every bit as spectacular and impressive as any of the dinosaurs. Indeed, animals larger than any of the dinosaurs—some species of whales—swim in today's oceans. While they, too, appeal to our sense of wonder, dinosaurs no doubt still have a stronger hold on our imagination.

CHANGED PERCEPTIONS

Our fascination with dinosaurs derives, at least in part, from the slowly unfolding history of their discovery and the gradual unearthing of the truths of their existence.

From the early years of the 19th century, museums began to fill with huge skeletons, carefully, but often wrongly,

THIS PAINTING *by Charles Wilson Peale (1741–1827) shows sophisticated technology being used in a fossil dig in the United States, in 1799.*

reconstructed from bones and other fossils recovered from sites all around the world. These remains of animals from a previously unrecorded past were brought to life again in petrified form. It was as if the fledgling science of paleontology had proved that the dragons and monsters of folklore and myth really did once exist.

In the 180 years or so since humans first learned of the existence of dinosaurs, our understanding of them and the worlds they inhabited has increased hugely. And with this growth in knowledge has come an increase in respect. We no longer think of dinosaurs in the way that earlier generations did—as

evolutionary mistakes that were selected for extinction to make way for more sophisticated animals. We now realize that they were an extremely successful and diverse group of animals that occupied all the lands of the world for 160 million years. The reign of humans on Earth pales in comparison with this long dominance.

DINOSAUR DIVERSITY

The diversity of dinosaurs as a group is now widely understood. Although a number of dinosaurs, including some of the best known, were huge, most were not, and many were small animals. In fact, the average size of a dinosaur was little bigger than that of a

modern sheep. Up to the present, researchers have named around 860 different types of dinosaurs, from bantam-sized insect-eating creatures to huge, lumbering beasts that weighed tens of tons. There were dinosaurs with horns, plates, spikes, clubs, frills, crests, and sails. Some had teeth like steak knives that could shred flesh, while others had batteries of flattened teeth for grinding plants to a paste. A number of dinosaurs had no teeth at all. Some were sleek and agile and probably capable of wolf-like speeds, while others were elephantine, with legs like tree-trunks that supported immense bodies.

The number of dinosaurs that have been discovered, studied, and named is now very extensive, and new discoveries are still coming to light. There are, however, many types of dinosaurs that

THE MODERN BLUE WHALE *is not only the largest animal on Earth, or in the water, it grows bigger than any animal, including the dinosaurs, that ever lived. These huge marine creatures can weigh more than 100 tons (98 t).*

we will probably never know about. All that we do know comes from fossil remains. Because fossils form only under certain conditions, and because many dinosaurs probably lived in environments where these conditions did not apply, no fossils have been found as clues to their existence.

PALEONTOLOGY

The science of paleontology is the study of fossils. Until relatively recently, the main focus of paleontological research into dinosaurs had been to work out how they were related to each other and where they fitted into the greater story of life. More recently, however,

paleontologists have been delving more deeply into questions of how dinosaurs lived, and just what were they like as living animals.

The fossil clues that paleontologists have to work with are surprisingly varied. Bones readily fossilize if the conditions are right, and sometimes complete dinosaur skeletons are found intact. On rare occasions, muscles, skin, and other soft tissue can also fossilize. Dinosaurs have also left their footprints, eggs, and dung, all of which provide clues about the animals that created them. Though far from complete, these discoveries have given us an ever-expanding picture of what the Earth and its inhabitants were like in the age of dinosaurs.

ONE OF THE LARGEST *of all the dinosaurs, Tyrannosaurus (left) dwarfed its contemporaries. Compsognathus (right), was a midget by comparison. It grew no taller than a modern chicken.*

TELLING GEOLOGICAL TIME

Rocks that have built up in layers over millions of years have provided the vital clues to the age of the Earth and its successive life forms.

Unraveling the history of the Earth has been the task of geologists who have been studying the rocks of the world for the past 250 years. They have built up a coherent story by examining how rocks form and how they have come to be placed in relation to each other.

LAYERS OF ROCK

The rocks of the Earth's surface are layered one on top of the other, like a stack of books. It follows, then, that the rocks situated at the bottom of the pile must have been laid down before the rocks closer to the top. By working on this principle, geologists are able to work out the relative ages of rocks by observing how the different types of rocks are arranged with respect to each other. By systematically mapping all the known rocks of the world and working out their relative positions, geologists have assembled what is known as the stratigraphic column. This is a register, or diagram, of the relative ages of different types of rocks.

A FOSSILIZED CRAB *(top left), found in Jurassic deposits in Australia.*

The stratigraphic column spans almost the full 4,600-million-year history of the world. This is a huge collection of information so, in order to make it easier to understand, geologists have divided it into a number of sections. These sections are what we call the geological eras, periods and epochs. For example, the Jurassic period is a section of the

DIFFERENT STRATA *of reddish sandstones, shales, mudstones, and siltstones can be distinguished in these spectacular cliffs in Arizona's Grand Canyon.*

CHARLES HAZELIUS STERNBERG led his family of three sons in a fossil-collecting business. At the feet of Charles Sternberg lie the fossil remains of Albertosaurus libratus, collected in 1917.

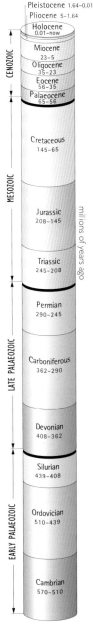

THIS
TIMELINE

*shows the order
and duration of
the different
geological eras
and periods in
the Earth's
history. The
dinosaurs lived
only during the
Mesozoic era,
appearing late
in the Triassic
period and
becoming
extinct at the
end of the
Cretaceous.*

stratigraphic column that is younger than the Triassic period that lies underneath it, and older than the Cretaceous period that sits above it. The Jurassic period represents a segment of time in the history of the Earth that began 208 million years ago and finished 145 million years ago. Periods are grouped into geological eras. For example, the Triassic, Jurassic, and Cretaceous periods are grouped together into the Mesozoic era. The periods in turn are divided into smaller segments known as epochs and stages.

Times are calculated by looking at the degree of decay that is displayed by particular elements found in certain types of rocks. The longer the rock has been in existence, the greater the decay that will have taken place. These techniques provide "absolute" time or age, which scientists measure in millions of years. When combined with information from the stratigraphic column, they allow us to give dates to other rock units. It comes as no surprise that the dates given

by absolute dating techniques agree with the order of dates given to us from the stratigraphic column.

FOSSILS IN ROCKS

The world has existed for 4,600 million years, and fossils have been found dating back 3,800 million years, but evidence for complex life forms does not appear until around 600 million years ago. Fossils dating back longer than this are relatively scarce.

However, a more abundant supply of more recent fossils has allowed scientists to plot the course of evolution and the development of life. As various types of animals and plant forms have come into existence and become extinct, they have left their fossils behind in the rocks within the stratigraphic column.

Because living things do change through time, their fossils can be used to date rocks and to make comparisons and correlations between rocks over large geographical areas. For example, dinosaur fossils are found only in rocks of the Mesozoic era, which extends from 245 to 65 million

years ago. Any rock in which a dinosaur fossil is found has to be from the Mesozoic era. A particular type of dinosaur, of course, is more restricted in its time range; identifying the dinosaur can help to narrow down even further the age of a kind of rock.

Unfortunately, however, dinosaur fossils are relatively rare. The shelly fossils, such as those of clams and snails, are more abundant and widespread, and are therefore more useful for this kind of dating.

THE FORMATION OF FOSSILS

Fossils are the vital clues on which the laborious but fascinating detective work of paleontology is based.

Fossils are records, written in stone or rock, of life through the ages. Anything that was once alive can leave fossil traces, but some organisms are better candidates for fossilization than others. Usually, only the hard parts of a plant or animal end up as fossils. Muscles, skin, and internal organs are rarely preserved. The shells of animals such as clams and snails and the bones of vertebrates are much more likely to be preserved than are the bodies of soft animals such as worms and jellyfish. Indeed, the fossil record of these creatures is almost nonexistent. In certain instances, footprints, eggs, and dung can also fossilize.

FOSSIL PROCESSES

There are a number of ways that fossils can form. Most fossils, however, involve watery environments and result from the burial of an organism's remains in the sediments of a river, lake, or sea. Once the soft tissues have rotted away, the bones or shell become encased in the surrounding muds and silts. As time passes, these sediments harden into rocks, and the bones or shells that are trapped within create an impression of their living

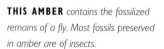

THIS AMBER *contains the fossilized remains of a fly. Most fossils preserved in amber are of insects.*

form. Sometimes the actual remains are totally replaced, cell by cell, with minerals that wash through the enveloping rock. This process is called "petrification." In other cases, the whole bone is dissolved, leaving behind a hole—a natural mold in the rock that can later fill up with minerals.

Other fossils are created in less usual ways. Insects and small animals can become trapped in tree sap that eventually hardens into the semiprecious stone amber and seals in a perfect copy of the entombed animal. Sometimes, the mineral silica

A FOSSILIZED TREE *forms part of a fossil forest at Lulworth Cove, in Dorset, UK. The doughnut shape is the fossilized remains of a collar of algae. These fossils are from the Jurassic—they are betwen 162 and 135 million years old.*

AMATEURS AT WORK

Not all fossils are found by professional paleontologists. Fossil collecting is an absorbing and popular hobby for many amateur enthusiasts, who sometimes make the important finds that lead to breakthrough discoveries. William Walker was a case in point. He was an amateur fossil collector who, in 1983, came upon an enormous fossil claw while fossicking in a quarry in Surrey, in southern England. He took his find to the British Museum of Natural History in London, where paleontologists soon realized that it belonged to a previously unknown dinosaur. A team of paleontologists went to the quarry and started digging. They eventually excavated a relatively complete skeleton of the new dinosaur, which they named *Baryonyx walkeri*. *Baryonyx* means "heavy claw," and *walkeri*, of course, is a tribute to the man who discovered it.

FISH *are the oldest vertebrates, but their fossils are relatively rare. Plant fossils are more common, especially in Carboniferous and Tertiary rocks. The palm and fish fossils (left) are from the Green River Formation in Wyoming, USA. The diagrams (below) show stages in the formation of dinosaur fossils.*

can fill the impressions in the rock left behind by an animal, resulting in a fossil shell or skeleton that glitters with the fire of precious opal. On rare occasions, the scalding ashes from a volcano can encase a hapless animal. The resulting fossil is a hole in the shape of the trapped creature.

By far the greatest number of fossils are the remains of shelled creatures that lived in shallow seas. Corals, clams, snails, and a host of other invertebrate animals make up the bulk of the world's fossil collection. More rarely, plants can become fossilized. Coal, for example, is the fossil remains of whole forests. Only rarely, though, do any traces of plant structure survive. These are destroyed as the coal is compressed to less than one-hundredth the thickness of the once-living plants that comprise it.

LAND ANIMALS

The remains of animals that once lived on land are fossilized even more rarely than those of plants. To be candidates for fossilization,

these animals need to have died close to or in a water course that floods and buries its prize in muds and silts. The fact that, at least as far as we know, all dinosaurs lived on land explains the scarcity of their fossil remains and makes it all the more likely that we will never know just how diverse they really were.

Animals that walked or ran across floodplains or tidal flats have left a record of their passing as fossilized footprints. Sometimes the trackways of whole herds of animals are preserved in this way. And in some cases, complete rookeries of dinosaur nests have been inundated and are now preserved as fossils.

Now let us turn to our richest geological museums, and what a paltry display we behold!

On the Origin of Species
CHARLES DARWIN (1809–1882),
English naturalist

BELOW THE SURFACE *of a lake, a dead dinosaur's flesh rots away or is eaten by aquatic creatures.*

LAYERS OF SILT *build up over the dinosaur's bones and prevent them from being washed away.*

WEIGHED DOWN *by sediment, the dinosaur bones are slowly replaced by minerals.*

MILLIONS OF YEARS *later, seismic disturbances bring the fossilized bones near the surface.*

DATING FOSSILS

The ability to date rocks and the fossils that are found inside them is of critical importance. Scientists have gradually devised a number of sophisticated ways to do so.

There are several methods of directly dating some types of rock. One —known as radiometric dating—measures the degree of decay of various isotopes (different forms of a chemical element) that are contained in particular minerals in the rock. An alternative technique, called paleomagnetic dating, involves the measurement of the ancient magnetism of a rock. Yet another, fission-track dating, examines the effects of uranium breakdown in zircon crystals.

RADIOMETRIC DATING

One of the most widely used radiometric dating techniques measures the rate of decay of the isotope potassium 40 to form the gas argon 40. Potassium is a common element in minerals such as feldspars, which are frequently present in basalt and other igneous rocks. A small proportion of naturally occurring potassium includes potassium 40, which decays at a steady, and known, rate to produce argon 40.

When a rock is molten, any accumulated argon gas can escape, and the "rock clock" is reset. Once the rock solidifies, the argon gas starts

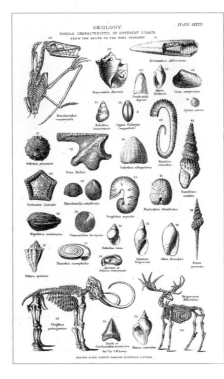

to build up again, trapped inside the rock. The longer the time since the rock solidified, the greater the amount of argon gas that accumulates. Measuring the amount of argon 40 gas relative to the amount of potassium 40 in a rock gives an accurate measure of how much time has elapsed since the rock solidified from its molten state.

Basalts are common rocks formed from lavas that spread across the landscape. With time they become interleaved with sedimentary rocks. Therefore, by having potassium/argon dates from basalts, it is possible to

BY DATING *different layers of rocks (top left) scientists can determine the age of fossils found in them. The 19th-century fossil illustrations (left) depict fossils from different periods, from the Jurassic to the post-Tertiary.*

determine the maximum and minimum ages of sedimentary rocks above and below them.

Potassium/argon dates are not the only isotope pairs that are used in radiometric dating. Other isotope pairs include rubidium/strontium, uranium/lead, and samarium/neodymium, but the basic technique is the same: By measuring how far these isotopes have decayed, an accurate assessment of how long the process has been going on for can be made. Different isotope pairs decay at different rates and are useful over different ranges of time.

One of the best known radiometric dating techniques is C14, or carbon dating. When an organism is alive, it takes in carbon 14, along with the rest of the carbon it needs from the environment in order to live. When that organism dies, it stops taking in carbon and the C14 that has accumulated inside begins to decay. As in other radiometric techniques, getting a C14 date involves

THE SHRIMP *(Sensitive High Resolution Ion Micro Probe) instrument at the Australian National University in Canberra is used to date rocks and forms of early life. It bombards zircon specimens with ions to release uranium and lead ions.*

measuring how much of the isotope is left in an organism. However, carbon 14 decays fairly rapidly and is only useful for dating once-living material less than 50,000 years old. It is, therefore, of no use in dating dinosaur fossils.

PALEOMAGNETIC DATING

For reasons that are not entirely clear, the magnetic poles of the Earth occasionally flip over so that magnetic north becomes magnetic south, and vice versa. A record of these magnetic flips is held within once-molten rocks such as basalt. Tiny magnetic particles within a lava line up with the prevailing magnetic field of the Earth and, when the lava solidifies as rock, the particles are set in their original orientation. Magnetic reversal events vary in length, and the complex history of magnetic reversals is preserved on the floors of the oceans. Study of the magnetic orientation of basalts on land can match them to the history from the sea floor, and an age of the basalt and its surrounding rocks can be determined.

FISSION-TRACK DATING

Uranium occurs naturally in the common mineral zircon. One of the isotopes of uranium, U238, is unstable and undergoes fission, a nuclear reaction where it splits and forms a more stable isotope. When this happens inside a zircon crystal, the parts of the isotope fly apart with such force that they leave minute scratches in the crystal matrix. Like all isotopic decay, the decay of uranium 238 occurs at a known rate, so that the longer a zircon crystal has existed, the greater the number of fission scratches, or tracks, that will be discernible. Essentially, then, fission-track dating consists of counting the fission scratches in zircon crystals.

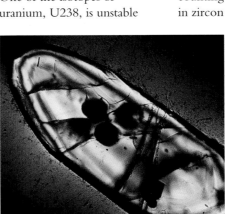

THIS ZIRCON CRYSTAL, *dated at 4,100 million years old, was found at Mount Narryer in Western Australia. The round holes are caused by the dating process undertaken by scientists at the Australian National University, in Canberra.*

IDENTIFYING AND CLASSIFYING DINOSAURS

Both living and extinct animals are classified by scientists

according to the similarities and differences between them.

Even before humans appeared on the scene, animals, in order to survive, had to develop some of the skills of the naturalist. They had, for example. to learn to classify the plants and animals around them into "safe" and "dangerous," "edible" and "inedible," and "approach" and "avoid." Basic classification, then, is fundamental to what we do, not only as humans, but as living beings.

TAXONOMY

Biologists use a method of classifying living things which reflects the relationships between them. This system of classification is known as taxonomy. It was developed by the 18th-century Swedish naturalist Carl Linnaeus (1707-1778), who published his ideas in a book, *Systema Natura*. Linnaeus's approach involved two revolutionary initiatives. First, he assigned to all known plants and animals a two-part name—the biological binomen. The first part of the name is the generic name—it identifies the genus. A genus is a collection of closely related species. The second part is the species name—a means of identifying a population, or group, of individual organisms that are distinguished by the fact that they can interbreed with each other but not with other living things. The second breakthrough by Linneaus was the idea of ranking similar organisms according to their degree of similarity. Similar species are grouped into a genus, similar genera are grouped into families, families are grouped into orders, and so on. The higher we go in this hierarchy, the less similarity there is between individual members of the group.

Eighty years after Linnaeus developed his system, Charles Darwin's theory of evolution would give added sense to theses classifications, by suggesting that organisms are related because they share a common descent. The more recently organisms have diverged from each other the more similar they appear.

THESE THEROPOD FOOTPRINTS, *recognizable by the imprint of three large toes, are preserved in sedimentary rocks at Kalkfeld in Namibia.*

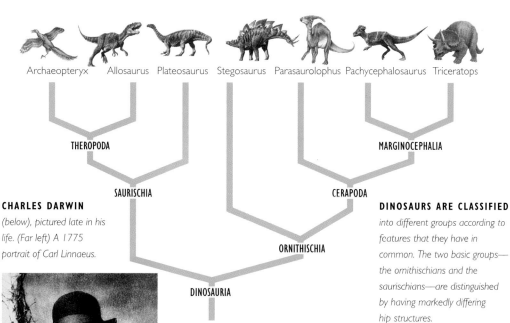

Archaeopteryx　Allosaurus　Plateosaurus　Stegosaurus　Parasaurolophus　Pachycephalosaurus　Triceratops

THEROPODA

MARGINOCEPHALIA

SAURISCHIA

CERAPODA

ORNITHISCHIA

DINOSAURIA

CHARLES DARWIN
*(below), pictured late in his
life. (Far left) A 1775
portrait of Carl Linnaeus.*

DINOSAURS ARE CLASSIFIED
*into different groups according to
features that they have in
common. The two basic groups—
the ornithischians and the
saurischians—are distinguished
by having markedly differing
hip structures.*

FOSSILS AND CLASSIFICATION

Classifying organisms known only from fossils presents some problems. First, we cannot demonstrate which individuals could mate with others. Therefore, the basis of the concept of "species" has to be changed. In paleontology, species are defined by fossils that are distinctly different in their structure from other fossils or living organisms.

Ranking fossils into genera, families, orders, and other groups depends on skeletal features that are shared with other fossils. The more features that two fossils share, the more closely related they are. Members of species are essentially identical. Members of the same genus share many features, while members of families share fewer.

A recent improvement on this system of classification is the recognition that evolutionary novelties can indicate relationship. A fish, a mouse, and a lizard all have bones, but this tells us little about how they are related. The fact that the mouse and the lizard both have four walking limbs tells us that they are more closely related to each other than either is to the fish, because the fossil record reveals that "legs" are a more recent evolutionary advance than "fins."

About a dozen unique characters identify dinosaurs. Many of these are subtle, but some are readily apparent. Unlike their predecessors, dinosaurs had a hip configuration that allowed the legs to move in a backward–forward motion under the animal. This conferred a number of advantages on dinosaurs, including a more energy-efficient way of moving and the possibility of attaining greater speeds. Another

common feature was a crest on the upper arm bone. This allowed the arms to be pulled together with greater power.

The classification of footprints, eggs, and other non-body fossils of dinosaurs presents other problems. It is only rarely that we can match footprints with a particular dinosaur. Eggs, too are difficult to classify unless the remains of embryos, juveniles, or adults are found in direct association. For the sake of simplicity, a separate set of names is given to footprint, egg, dung, and other trace fossils. It is quite possible that three different names have been given to the same dinosaur's bodily fossils, its eggs, and its footprints.

A NEST *of oviraptorid eggs, found in the Gobi Desert in Mongolia. Eggs are often hard to match with the dinosaur that laid them.*

WHAT DINOSAUR FOSSILS CAN TELL US

Everything we know, or can conjecture, about dinosaurs

and how they lived is based on what fossil remains can reveal.

Fossils are the remains of once-living organisms. It is the paleontologist's job to reconstruct the life of extinct animals based on the fossils they have left behind. Dinosaur fossils are among the most intensively studied fossils of all.

BONES AND SKELETONS

The most common dinosaur fossils are individual bones and, more rarely, complete skeletons. From these finds, paleontologists are able to construct theories and draw conclusions about the evolution of dinosaurs— what different species were like in life, and how they were related to each other and to other creatures that lived at the same time.

Although rare, skeletons present a more complete picture than separate bones. Complete skeletons have been found of only very few dinosaurs. However, most of the major groups of dinosaurs include at least one member for which a skeleton is known. Skeletons help us to construct the missing pieces of other dinosaurs that are known only from fragmentary remains, or even single bones.

Reconstructing any ancient creature from a collection of fossil remains is a difficult task. It is essential to combine only like with like, and not to

THESE HADROSAUR BONES, *dating back to the late Cretaceous, are in the Dinosaur Provincial Park in Alberta, Canada.*

generalize from one type of dinosaur to another. For example, to combine an isolated plate from an armored dinosaur with the shin bone of a small carnivore and the ribs of a long-necked herbivore would result in a strange hybrid that resembled no dinosaur that ever lived. Such mismatching, though, has sometimes occurred.

FEATURES IN COMMON

Bones and other fossil remains paint a cumulative picture of the evolutionary changes that dinosaurs experienced over time. Coupled with evidence of age, gained from the radiometric dating of rocks, changes of bone structure can be used to reconstruct dinosaurs' evolutionary history.

The starting point of this reconstruction is the phylogenetic analysis. Paleontologists carefully note individual features of the bones of different dinosaurs and study how these features are distributed. The more closely two dinosaurs are related, the more features their bones should have in common. Of particular interest are features that are evolutionary novelties—new features not seen in the ancestors. For example, a crest on the arm bone (the delta pectoral crest) is found in all dinosaurs but in no other group of animals. This suggests that dinosaurs are more closely related to each other than any of them is to their immediate ancestors or to other animals that lack this crest. The more features we find that have a similar distribution, the more faith we can place in the groupings of dinosaurs.

A REBUILDING PROCESS

Bones and skeletons are also our starting point in rebuilding the the appearance of an extinct creature. To begin with, the size of a bone will reflect the size of its original owner. The paleontologist must be conscious that the features of these bones are clues to the ways in which they served a living animal. For example, bones from the "hands" of a dinosaur that walked on all fours will be more robust than the same bones of a relative that walked with its hands free. Again, sharp claws are

for slashing or holding, blunt claws are used for walking on. Meat-eaters require sharp teeth for ripping flesh; plant-eaters need grinding teeth to pulp their food.

Bones also give important clues about the soft parts of a dinosaur that have not fossilized. Muscles attach to bones and leave scars. Careful study of the size and position of these scars can reveal the way in which the animal moved a limb and how strong it was. Dinosaurs' brains were encased in bone, so, even though the brain is never fossilized, the

THIS FEATHER FOSSIL, *dating back to the early Tertiary, was found in the Green River Formation, Wyoming, USA.*

space in which it was housed can give a good idea of what the living brain was like. Nerves and blood vessels coursed through various bones, leaving behind holes that tell of their passage.

Diseases and injuries suffered by an animal during life can also be reflected in bone fossils. An extensive list of diseases—including various cancers, arthritis, and gout—have been identified in dinosaurs from bone fossil evidence. Injuries detected include broken bones and gouges from fighting with members of their own species or with predators.

There is also a lot to be learned from the arrangement of bones where they are discovered. An intact skeleton reveals a quick burial after death, whereas a scattered collection of bones may indicate that the carcass was scavenged. Isolated bones may have been carried off by a predator or washed away from the rest of the skeleton in a stream or flood.

CLEAR IMPRESSIONS *of the skin and bony tendons of a hadrosaur from the late Cretaceous are preserved in this fossil (left) from Utah. The cell structure of a fossilized Jurassic dinosaur bone (below) is visible in this specimen from the Morrison Formation in Colorado, USA.*

RARER FOSSILS

Many important insights into how dinosaurs lived can be gleaned from non-bone fossils such as eggs, dung, footprints, and trackways.

The rarest of dinosaur fossils preserve soft tissue—feathers, skin, muscle, or, very rarely, internal organs. These delicate structures are easily destroyed and exceptional circumstances are required for them to fossilize.

Feathers are most famously known from some of the specimens of *Archaeopteryx*, which was first discovered in 1860. More recently, however, a suite of early birds and birdlike dinosaurs have been found in China with feathers and featherlike structures still attached. In these cases, the carcasses of the animals were preserved in extremely fine-grained sediments and were unaffected by scavengers. Perhaps after they died, their carcasses fell to the beds of lakes or seas where there was little or no oxygen, thus preventing scavengers from reaching the bodies. The occurrence of feathers associated with the skeletons of small theropod dinosaurs is crucial to our understanding of the origin of birds.

Skin, or impressions of skin, has been found with several dinosaurs. It reveals that most had a scaly covering, similar to that of crocodiles or of some lizards. Skin impressions are preserved in cases where a dinosaur fell

By this theory innumerable transitional forms must have existed.

On the Origin of Species,
CHARLES DARWIN (1809–1882),
English naturalist

into a fine mud that hardened soon after contact.

The small theropod *Scipionyx*, from Italy, provides the best example of preserved internal dinosaur organs. In this exceptional juvenile specimen, the liver, intestine, and various muscles have been preserved. Some specimens from Liaoning, in north-eastern China, also retain some internal organs, and a number of skeletons, such as the remains of *Seismosaurus*, retain small stones, called gastroliths, that were swallowed by the animal. These indicate the shapes and location of parts of the digestive system.

NESTS, EGGS, AND EMBRYOS

While fossils of dinosaur eggs and embryos have been known for over 100 years, it is only within the last couple of decades that they have been studied in detail. They have revealed a wealth of information about dinosaur reproduction, development, and behavior.

Whole rookeries of dinosaur nests—found in the United States, Argentina, and Mongolia—show that some dinosaurs nested in huge groups and that they returned to their nesting grounds year after year.

In many surprising ways, dinosaur nesting behavior resembles that of modern birds. Spacing of nests, for example, is determined by the length of the adult dinosaur. This reflects the way that nest spacing in modern communal nesting birds is set by the size of the adult bird. Studies of dinosaur nests reveal that the young of some dinosaurs stayed in the nest for a certain time after hatching, and that parents must have tended to them and fed them during this period. Other types of dinosaurs appear not to have given parental care to their offspring. They seem to have deserted the eggs as soon as they

SINORNITHOSAURUS,
an early Cretaceous dinosaur, was discovered in Liaoning Province, north-eastern China, in 1999. This fossil showed evidence of featherlike structures on the body.

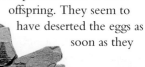

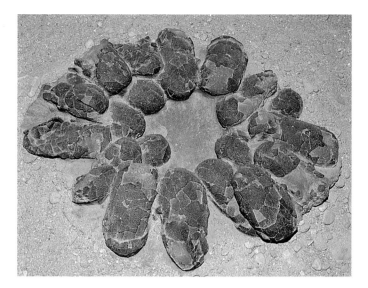

SAUROPOD TRACKMARKS
These herbivores walked on four feet.
The large prints are from the back feet.

IN THE 1920S, *a number of nests, more than 70 million years old, of Oviraptor eggs were discovered in Mongolia. Many of the eggs were still intact and some contained the fossilized remains of tiny embryos.*

SMALL THEROPOD TRACKMARKS
These small carnivores left delicate, birdlike tracks.

were laid, leaving the young to fend for themselves.

OTHER FOSSILS

Some of the most intriguing dinosaur fossils are the remains of dung. These fossils, known as coprolites, can show what dinosaurs ate, and can give an idea of the size of the anus and the structure of the lower digestive tract. However, coprolites provide little direct evidence that links them to the dinosaurs that produced them. While there are analyses that suggest which dung belongs to which dinosaur, it is unlikely that these identifications can ever be positively confirmed. Another feature of coprolites is that they are frequently associated with dung beetles that burrowed under the dung and filled their burrows with dinosaur feces.

Footprints and trackways have been found for most of the major dinosaur groups, and they provide important

clues to the movement and behavior patterns of their makers. The spacing between footprints in a trackway indicates both the size of the animal and the speed at which it was traveling. A number of trackways at a site that point in a similar direction suggests that some dinosaurs moved in groups. In those rare sites where multiple trackways exist, the composition and structure of a herd can be determined. In some cases a dramatic tableau of dinosaur life has been preserved—showing, for example, evidence of a predator scattering a herd of smaller dinosaurs or stalking larger prey. However, associating footprints and trackways with the dinosaurs that made them is an imprecise science.

CERATOPSIAN TRACKMARKS
These herbivores walked on four legs.
The front feet made the smaller prints.

LARGE THEROPOD TRACKMARKS
These large carnivores were bipedal.
Their back feet each had three toes.

DINOSAUR EGGS *were small in relation to the animals' body size. This is because larger eggs would have had shells too thick for hatchlings to break.*

Oviraptor's egg

Emu's egg

Possible ornithischian egg

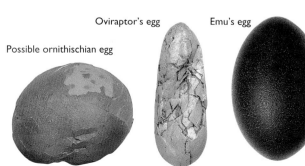

EARLY DINOSAUR DISCOVERIES

For centuries, people have been finding dinosaur fossils.

It is only in the last 150 years or so, however, that scientists

realized that dinosaurs formed a distinct group.

We will never know who discovered the first dinosaur fossil, or even when, or where, it was found. We do know that Western scientific attention to this group dates back to the 17th century, but at that time no one would have thought that the bones they were studying belonged to giant extinct reptiles.

PLOT'S DISCOVERY
The first description of a dinosaur bone that we know about was published by an Englishman, Robert Plot, in 1677. The bone Plot described is almost certainly

SKELETONS OF IGUANODON on display in the Royal Museum of Natural History in Brussels. In 1878, a wealth of Iguanodon skeletons came to light in a coal mine near Bernissart, Belgium.

the base of a theropod femur, which was possibly from *Megalosaurus* or a similar dinosaur. Plot, clearly puzzled by the find, described it merely as an "enigmatic thighbone" and assumed that it belonged to a giant human from before the time of the flood described in the biblical Book of Genesis. In Plot's day, scientists did not yet recognize that species could become extinct. For them, whatever was found from the past must represent a form of life that still existed. However, as more and more strange and baffling animal remains were exhumed all over Europe, and as more of the world was explored, it became increasingly difficult to explain them away simply as antediluvian giants. By the early

1800s, the idea that species could become extinct had become widely accepted, at least among members of the scientific community.

DINOSAURS RECOGNIZED
Thoughout the 18th century, there are references in the scientific literature to what were probably dinosaur bones, and dinosaur footprints, though not recognized as such, were known in New England at that time.

A major breakthrough in the understanding of dinosaur fossils occurred about 1818, when the English naturalist

DEAN WILLIAM BUCKLAND (1784–1856), shown here standing at the dinner table, was Professor of Geology at the University of Oxford and Dean of Christ Church in England. His eccentric household comprised a veritable menagerie of animal species.

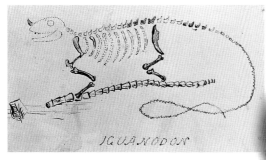

IGUANODON *was named by Gideon Mantell (1790–1852). While the name remains unchanged, Mantell's description of the animal has been proved inaccurate. His sketch of the newly discovered dinosaur is shown at left.*

Dean William Buckland began a systematic survey of animal fossils from all over England. Among them were some fragmentary bones from Oxfordshire, including a piece of a jaw with teeth. In 1822, Buckland named the beast from which these came *Megalosaurus* (meaning "giant lizard"). This represented a pioneering move toward the recognition of dinosaurs as extinct reptiles.

Soon afterwards, in 1825, the English dentist Gideon Mantell reported on some fossils which were discovered, probably by his wife, Mary, near the town of Cuckfield in Sussex. These finds included a conical bone and some teeth that, as Mantell observed, closely resembled those of a living iguana. Mantell named the disovery *Iguanodon*, and concluded that this now extinct animal was a giant variant on the living form.

Both *Iguanodon* and *Megalosaurus* were, at the time, assumed to be large quadrupeds and early reconstructions represented them as such. This assumption, though it seems quaint to us now, was reasonable when we consider that no bipedal reptiles were then known.

New dinosaurs continued to be discovered in Britain and other places, but it was not until the 1840s that the great English anatomist and paleontologist Sir Richard Owen recognized that all these animals shared certain distinctive features and that therefore they all belonged together as a group. Finally, in 1849, he grouped them together, in the class name of "Dinosauria." Thus, the word "dinosaur," which means "terrible lizard," became part of the English language.

FURTHER INSIGHTS

In 1858, some fossil bones came to light in the Cretaceous marls of New Jersey. Joseph Leidy, a professor of anatomy in Philadelphia, assembled them into a relatively complete skeleton of a new dinosaur that he named *Hadrosaurus*. With so much material at his disposal, Leidy could observe features of this dinosaur that had never previously been guessed at. For one thing, the disproportionate development of the back legs, as compared to the front ones, suggested to him that this huge creature walked on its back legs.

Twenty years later, a Belgian coal mine near the village of Bernissart yielded the partial and complete skeletons of almost 40 *Iguanodon* individuals. These finds became the responsibility of Louis Dollo (1857-1931) of the Brussels Royal Museum of Natural History. Over a number of years, Dollo was able to reconstruct many *Iguanodon* skeletons, in lifelike pose. The finds revolutionized our understanding of dinosaurs and proved, among other things, that some species of dinosaurs lived in groups rather than as solitary animals.

CLAW OR HORN?

Louis Dollo's reconstructed skeletons of *Iguanodon* finally laid to rest a misunderstanding that had lasted for more than half a century. Gideon Mantell had concluded that the conical bone belonging to *Iguanodon* was a horn at the tip of its snout. Given the assumption he made, this seemed reasonable, as some living iguanas do, in fact, have short horns on their snouts. During the 1850s, Richard Owen actually reconstructed *Iguanodon* as a large, lizardlike, "horned" quadruped. Thanks to the Bernissart discoveries and Dollo's subsequent work, we now know that *Iguanodon* was a bipedal herbivore. The conical bone was a deadly thumb spike that the animal probably used for defense.

RICHARD OWEN'S *greatly flawed 1854 reconstruction of* Iguanodon *showed it as a solidly built quadruped with a rhinoceros-like horn.*

DINOSAUR HUNTERS OF THE PAST

A frenzy of fossil hunting that began in the 1870s led to a spate of exciting dinosaur discoveries.

EDWARD DRINKER COPE *probably named more reptiles, both living and extinct, than any other researcher.*

In 1859, Charles Darwin published his *On the Origin of Species*, in which he propounded his theory of evolution—the idea that life forms had changed through time and that all living things were related by descent. As the theory of evolution became widely accepted, at least among members of the scientific community, in the late 1800s—and the findings of paleontologists were seen as the most compelling support for it—the race to find and study dinosaurs became a major focus of scientific endeavor, especially in the United States.

AN INTENSE RIVALRY

The competition to find and name new dinosaurs led to an acrimonious rivalry between two greats of American paleontology. Othniel Charles Marsh, of Yale University, and Edward Drinker Cope,

of the Academy of Natural Sciences in Philadelphia, started their professional careers as friends, but soon they were bitterly competing against each other in the search for and study of new fossils. Both independently wealthy, they funded their expeditions into the American West. In some cases, they paid off other paleontologists, literally turning up at their sites, purchasing their fossils, and taking them back to the East Coast for study.

The two men even bought fossils from under each other's noses and employed spies in one another's camps. In their frenzied attempts to beat each other to the newspapers with new descriptions, they often telegraphed brief, sketchy, and sometimes misleading

accounts of their finds from the field. They frequently ventured into areas that were still considered dangerous frontiers. Within days of the Battle of Little Big Horn, in June 1876, for example, Cope and his team were hard at work excavating fossils just a few miles away from the site of the recent conflict.

The results of this intense feud were mixed. On the positive side, many of the now more familiar dinosaurs—such as *Apatosaurus, Diplodocus, Stegosaurus, Allosaurus,* and *Triceratops*— were revealed to the world by Cope and Marsh parties between 1870 and 1890. But on the negative side of the ledger, the result of their very

OTHNIEL CHARLES MARSH *is pictured here with a team of fossil hunters. Marsh is standing in the center of the back row. Marsh and his arch-rival Edward Drinker Cope between them named more than 130 dinosaurs, including most famous North American species.*

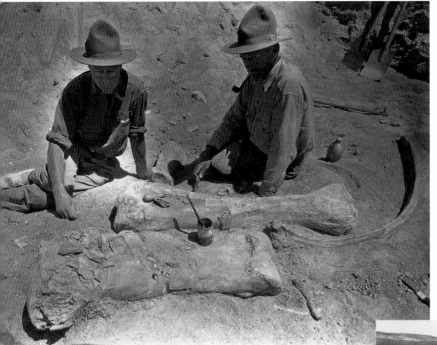

undignified public feuding was the decision made by the United States government to withhold public funding of paleontology for 25 years. There is no doubt, however, that the discoveries that were made by Cope and Marsh laid the groundwork for future exploration in some of the most important fossil-bearing localities in North America.

OTHER AMERICANS

Others carried on the work begun by Cope and Marsh in North America. The field work of Barnum Brown and Henry Fairfield Osborn of the American Museum of Natural History was of considerable significance. Among the finds for which they are remembered is that of *Tyrannosaurus rex*, which they found in Montana in 1905. Brown and Osborn worked mainly in the western United States, but they also collected in western Canada, especially along Red Deer River, in what is now Alberta. These expeditions were conducted by flatboat.

After Brown and Osborn, Charles H. Sternberg—whose passion for fossil collecting dated back to childhood—and two of his sons, Charles M. and Levi, continued to scout the late Cretaceous deposits in Alberta. Among the dinosaurs discovered by the Sternbergs were *Albertosaurus, Edmontonia, Styracosaurus* and *Lambeosaurus*.

OUTSIDE AMERICA

The rush of discoveries that began in about 1870 with the expeditions of Cope and Marsh is often regarded as a first golden age of dinosaur paleontology. The majority of its early finds were made in North America. This changed in 1907 when giant dinosaur bones were discovered at Tendaguru, in what is now Tanzania. Dr. Werner Janensch, curator of fossils at the Berlin Museum, was sent to Tendaguru, where he assembled a huge field party of almost 500 people. Many fine specimens were found, including the massive skeleton of *Giraffatitan* that is still on display in the Berlin Museum.

Among the last discoveries of this first golden age were those made in Mongolia and China. Beginning in 1922, the first motorized expeditions, led by Roy Chapman Andrews, were sent to central Asia by Osborn and the American Museum of Natural History to search, not for dinosaurs, but for evidence of the oldest humans. Although these expeditions failed in their primary task, they yielded a wealth of dinosaur finds, including the first dinosaur eggs and remains of *Oviraptor, Protoceratops*, and *Velociraptor*.

This first golden age ended in the 1930s. After that, and although dinosaur exploration continued, paleontologists became preoccupied with other issues. Another golden age, which is still with us, commenced in the 1970s.

CHAPTER TWO

THE TRIASSIC PERIOD

The first dinosaurs appeared sometime during the late

Triassic period, roughly 228 million years ago.

The environment in which the first dinosaurs lived would be unrecognizable to us. It was a world in which there was only one major landmass, where there were dry, red landscapes and forests without a single flower.

Animals at one end of the world would have looked more or less like animals at the other, because there were no seas to stop them spreading across the globe. But by the end of the Triassic, many features of today's world were in place: Forces were tearing the supercontinent apart, and more familiar plants and animals were appearing. As well, significant variations were developing in animals and plants found in different places. By the end of the Mesozoic, there would be a world that was outwardly similar to ours, but much of what made it look "modern" got its start in the Triassic.

Since the 1960s, plate tectonics has been one of the primary paradigms in geology. According to plate tectonics, the crust of the Earth is divided into broad, rigid, but constantly moving, plates that "float" on the more plastic layers of the Earth's mantle. Wherever plates interact, we find intense geological activity—earthquakes, growing mountains, and volcanoes. As a result of

plate movement, the Earth's continents are constantly in motion, a phenomenon known as continental drift.

PANGEA, THE SUPERCONTINENT

At various times in the past, all the Earth's landmasses have been joined. One of those periods spans the boundary between the Paleozoic and Mesozoic eras, when today's continents were joined in the single landmass of Pangea. North America, Europe, and much of Asia formed the northern part of the supercontinent, and the southern part consisted of South America, Australia, Africa, India, and Antarctica. At the beginning of the Triassic period, Pangea

THE TRIASSIC LANDSCAPE *was mainly arid, but, especially in regions near water, there were richly vegetated pockets where plant life—including conifers, gingkoes, ferns, and cycads—was able to thrive.*

extended from the South Pole to the middle of the Northern Hemisphere, and it gradually drifted northward. At the end of the Triassic, Pangea was centered on the equator. Consequently, climates changed, growing gradually drier and warmer.

Scientists agree that global climates were seasonal during the second half of the Triassic, with alternating warm-cool and wet-dry cycles. Worldwide climate was probably warmer and drier than today, with a broad arid belt in the middle of Pangea and more humid conditions toward the northern and southern ends. Many late Triassic rocks represent ancient sand dunes, and they tend to be red as a result of oxidized iron—an indication of arid conditions, according to some paleoclimatologists. Evaporites—minerals formed when salty water evaporates—were common in the late Triassic.

VEGETATION

Needle-trees, similar to living Norfolk Island pines, and monkey-puzzle trees, some of which were very tall, grew during the Triassic. We can see remains of these conifer forests in places such as

WASSON BLUFF *(above), in south-eastern Canada, was the site of important finds of prosauropod fossils in the late 1990s. (Right) Erosion pins are installed at Wasson Bluff in June 1999.*

Petrified Forest National Park in Arizona. Ferns, including tree ferns, were common and diverse; and cycads—thick-leaved plants that outwardly resembled palms—were also common. However, there were no flowering plants.

FAUNA

When continents collide, giant mountain ranges arise. This is still happening today in the Himalayas, where India is pushing into the rest of Asia. It also occurred when Pangea formed, but by the time the dinosaurs appeared, these mountain ranges were old and weathered down. As a result, and also because climates were broadly uni-form throughout Pangea, there was little to prevent animals from spreading far and wide. If you could go back to Pangea in the late Triassic you would see a range of animal life. On land, there would be herds of small, usually bipedal dinosaurs, as well as small, shrewlike early relatives of mammals and crocodiles that looked more like reptilian wolves. In the water, you would see various crocodile-like animals, among them large amphibians and distant relatives of true crocodiles.

Later in the Mesozoic, as Pangea gradually broke into separate landmasses—the result of plate tectonics—more diverse forms of animal life began to appear.

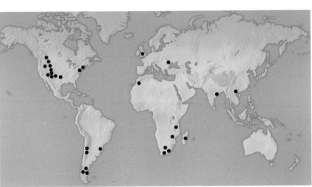

THE SUPERCONTINENT *Pangea (above) was the Earth's only large landmass during the Triassic, although forces that would break it up were already at work. Sites where remains of Triassic dinosaurs have been found (left).*

THE JURASSIC PERIOD

Throughout the Jurassic, changes occurred that resulted in a much greater diversity in landforms, flora, and fauna.

Changes occurred very gradually during the Jurassic. In the early Jurassic, world conditions were very similar to those of the late Triassic. In fact, some scientists make no distinction between these periods, often discussing them as though they were one. In the early Jurassic, the world was still generally warm and arid, especially near the equator, and there was little difference in land animals from one region to another.

THE JURASSIC LANDSCAPE *was much less arid than the Triassic one, thanks to generally warmer and more humid climatic conditions. More expansive forests grew, and plant groups that were present in the Triassic diversified greatly.*

DRIFTING APART

Change, however, was under way. At the very end of the Triassic, the slow but inexorable breakup of Pangea started. The ancestral Atlantic Ocean began to form as North America separated from Europe and Africa. If you visit Dinosaur State Park in Connecticut, you can see an extraordinary sight, thousands of footprints made by dinosaurs that lived in an early side valley of what would become the ancestral Atlantic. By the end of the Jurassic, we see separations, not only between the Americas and the Old World, but also between the world's northern and southern landmasses. What would later become the southern continents—South America, Australia, Africa, Antarctica, and India—were to remain in contact for the rest of the Jurassic and for much of the Cretaceous. The broad landmass they formed is known as Gondwana. The northern part of the former Pangea is now called Laurasia.

EXPANDING OCEANS

Scientists generally agree that both the North and South poles were free of ice throughout the Jurassic. It also seems probable that during periods of increased tectonic activity there was some slight expansion in portions of the oceanic crust. So, the lack of polar ice increased the amount of water in the

DURING THE JURASSIC, *Pangea slowly broke up to form the northern continent of Laurasia and the southern continent of Gondwana (above). Sites of finds of Jurassic dinosaur fossils (left).*

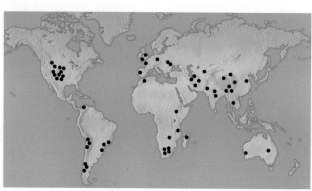

THE SEA *has eroded joints in the hard Jurassic rock in this cliff face, adjacent to Lulworth Cove, in Dorset, southern England, and then eroded softer rock behind, to create the formation known as the Stairhole.*

JURASSIC DINOSAURS

Jurassic conditions were conducive to the spread of reptiles, and many species of dinosaurs appeared during this period. By the end of the Jurassic, marked differences were beginning to develop between species inhabiting the northern and southern continents.

The late Jurassic, however, was predominantly the era of the giant sauropods, and it was at this time that they achieved their greatest diversity. Other plant-eating dinosaur groups—including ornithopods and, in some places, stegosaurs—were also present. Theropods, too, continued to diversify. There is reason to suspect that the fossil record of smaller theropods from the Jurassic does not truly reflect either their diversity or their numbers.

oceans, and the expanding sea floor dispersed the water more widely. As a result, sea levels around the world rose during the Jurassic, and large sections of the continents were flooded. This in turn added to the breakup of landmasses that tectonic activity was causing; expanses of water now separated what had been continuous stretches of land.

WARM AND HUMID

The world was uniformly warm during the Jurassic, although slightly cooler than during the Triassic. In contrast to the variable Triassic seasons, seasons during the Jurassic probably varied only slightly. However, after the early Jurassic, climatic differences probably became more accentuated, largely because of the increased fragmentation of landmasses. Everywhere was warm, but some areas became more humid and received greater rainfall than others.

There were probably still no flowering plants during the Jurassic. Although there is evidence to suggest that the group that included flowering plants was present at this time, it seems that these did not evolve flowers until the late Cretaceous. Forests, however, flourished and spread in the warm, wet Jurassic climate. They continued to be dominated by needle-bearing conifers, cycads, tree ferns, and ginkgoes.

THE CRETACEOUS PERIOD

The Cretaceous period, which began 144 million years ago and lasted for 80 million years, forms a bridge in time between the earlier Mesozoic and the Cenozoic, the age of mammals.

By the beginning of the Cretaceous period, the Earth was beginning to take on many of the features that are familiar to us today. As the continents continued to move farther apart, groups of plants and animals took on more regional characteristics. The fragmentation of landmasses also had an effect on the climate. As the Cretaceous progressed, climates around the globe gradually became more seasonal, and the annual variation in rainfall and temperature slowly became more pronounced.

THE FIRST FLOWERS

The most significant biotic change that occurred during the Cretaceous was the emergence of flowering plants. They first appeared early in the period and were more abundant in disturbed habitats where dominant conifers and cycads had been removed. By the end of the Cretaceous, there was a great diversity of flowering plants, and some familiar-looking groups, such as the water lilies, magnolias, and sycamores, had appeared. However, there were still no grasses—they did not arrive until the Cenozoic.

The arrival of flowering plants had a cascading effect on all other groups of organisms. There is evidence that complex, mutually advantageous relationships existed early in the period between plants and insects. Insects such as colonial bees, for example, relied on the plants for food, while the plants in turn relied on the insects for pollination.

A DIVERSITY OF DINOSAURS

There was also a significant shift in the nature of plant-eating dinosaur faunas in some parts of the world. Long-necked sauropods continued to be diverse early in the Cretaceous, but the numbers of these huge dinosaurs declined later throughout the period. By the later Cretaceous, the dominance in the northern continents had shifted to the ornithopods and ceratopsians. This latter group of horned animals was among the last groups of dinosaurs to evolve.

FLOWERING PLANTS *appeared for the first time during the Cretaceous period. Plant-eating dinosaurs cleared areas of conifers and cycads, creating spaces that were conducive to the development of these plants.*

Some scientists believe that the changes in the dinosaur fauna, and its increasing variety, were a response to the expanding diversity of flowering plants in the local flora.

FLUCTUATING SEAS

Sea levels throughout the world rose and fell a number of times during the latter half of this period. North America was essentially cut in half by a shallow sea that extended from present-day Hudson Bay to the Gulf of Mexico. Western North America was connected to eastern Asia, and some groups of dinosaurs—such as the ceratopsians, the tyrannosaurids, and the pachycephalosaurids—were unique to these two regions. Today's landmasses of South America, Antarctica, Madagascar, and India may have been closely linked, and all shared several unique groups of dinosaurs—such as the titanosaurid sauropods and abelisaurid theropods—during the Cretaceous.

Dinosaurs are also known to have existed in Australia during the Cretaceous. At that time, Australia was much

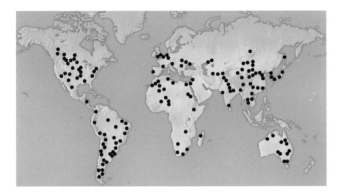

THE CONTINUING BREAKUP *of the world's landmasses during the Cretaceous (below) led to a greater diversification of species. Cretaceous dinosaur sites (left) are more numerous than those from the two preceding periods.*

closer to the South Pole than it is today and would have experienced extended periods of darkness, as Antarctica still does. Some of the Australian Cretaceous dinosaurs, most notably *Leaellynasaura*, had enlarged orbits, which suggests that they had large eyes that were specially adapted for night vision.

THE SEVEN SISTERS CLIFFS *in East Sussex, southern England, are typical of the white chalk cliffs that were formed during the Cretaceous period.*

A CATACLYSMIC EVENT

At the end of the Cretaceous period, important changes were occurring around the globe that would dramatically affect life on Earth. Sea levels were falling and temperatures everywhere were dropping slightly. Some parts of the world were experiencing extensive volcanic activity. Perhaps most significant of all, however, was a single extra-terrestrial cataclysmic event— a large asteroid or comet hit the Earth, with devastating consequences.

Any or all of these pheno-mena may have played a role in the major extinctions that occurred at the end of the Cretaceous, in which more than half the world's animals, including all of the Mesozoic dinosaurs, died out.

DINOSAUR ORIGINS

Sometime during the middle Triassic, one of the most successful archosaur lineages appeared—the dinosaurs.

Archosaurs are a group of animals that included the last common ancestors of living birds and living crocodilians. They were present in the early Triassic, but it was not until the middle of the period that they began to dominate the Earth's dry-land habitats.

The first archosaur would have looked like a strange cross between a crocodile and a dog. Although the ancestral archosaur remains undiscovered, we can get a glimpse of what it probably looked like with *Euparkeria*, a small animal known from the Triassic of South Africa. *Euparkeria* measured about 20 inches (50 cm) from tip to tip. We believe it was a predatory animal, with sharp teeth and powerful jaw muscles.

Although not warm-blooded in the sense that birds and mammals are, the earliest archosaurs were probably more agile and active than other reptiles of the era. Like modern crocodilians and birds, they probably had a four-chambered heart and the ability to drag their limbs under their body while walking.

AMMONITES were shelled marine carnivores of varying shapes and sizes. They evolved more than 500 million years ago, and became extinct, along with the dinosaurs, at the end of the Cretaceous.

ARCHOSAUR GROUPS

Early in their biological history, the archosaurs split into two groups. In one of these groups, the quadrupedal nature of the early archosaurs remained dominant. Some of these animals grew to be quite large, and may have been the dominant predators in some places during the middle Triassic. Later, the group became dominated by animals that were semiaquatic—these were the ancestors of the crocodiles and alligators that we see today.

The other group adopted a different approach toward predation. The members of this group remained small, but the length of the hindlimb increased. Eventually, some members became bipedal, walking mainly on the hind-limbs and using the forelimbs for grasping. The ancestral dinosaur was such an animal.

Although we have no fossils from the true ancestor of the dinosaurs, primitive dinosaurs, such as *Eoraptor*, can give us

STAGES in the development of a bipedal stance.
❶ The ancestors of dinosaurs sprawled like lizards.
❷ Some reptiles, such as crocodiles, can pull their limbs under their body.
❸ All dinosaurs could stand erect, supported by legs tucked beneath the body.

some idea of what it was like. The first dinosaur was probably no larger than a sheep. It was most likely predatory. It differed from other reptiles both in its bipedal stance and in its upright posture. Most living reptiles "sprawl"—their limbs project from the sides of the body, and their bellies are close to the ground when they walk. As they move, the body is thrown into a series of S-shaped curves.

Early archosaurs could, to some extent, draw their limbs into a vertical posture, but dinosaurs evolved their hind-limbs into a bipedal stance. The head of the thighbone (femur) was turned inward, so that the shaft projected downward, not outward, from the hip socket. Because of this, the hip structure evolved in a particular way. In other reptiles, the hip socket is a solid cup, because the femur is pulled inward by the hip muscles.

In the case of dinosaurs, the femur is pulled upward by the hip muscles. As a result, the hip socket is open and bears a prominent shelf of bone along the top rim. The belly was kept far from the ground, and the trunk remained straight when the animal moved.

This development is of vital importance. In reptiles such as lizards the muscles that move the limbs also control breathing—lizards cannot breathe and walk at the same time, and the distance they can run without having to stop is very limited. But reorientation of the hindlimb in dinosaurs meant that different muscles controlled walking and breathing, so that dinosaurs could do both

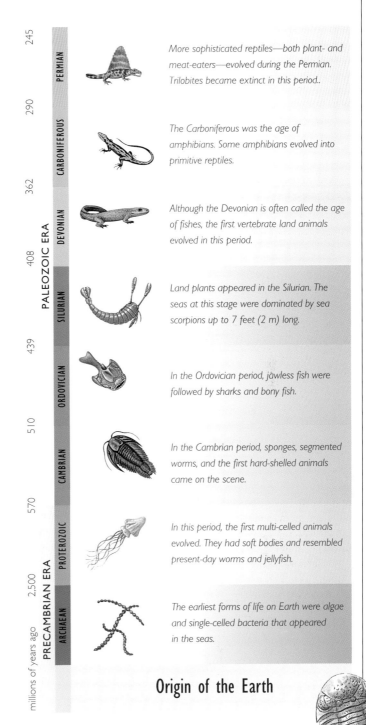

millions of years ago		
	PERMIAN	245
	CARBONIFEROUS	290
	DEVONIAN	362
PALEOZOIC ERA		408
	SILURIAN	439
	ORDOVICIAN	510
	CAMBRIAN	570
PRECAMBRIAN ERA	**PROTEROZOIC**	2,500
	ARCHAEAN	

More sophisticated reptiles—both plant- and meat-eaters—evolved during the Permian. Trilobites became extinct in this period..

The Carboniferous was the age of amphibians. Some amphibians evolved into primitive reptiles.

Although the Devonian is often called the age of fishes, the first vertebrate land animals evolved in this period.

Land plants appeared in the Silurian. The seas at this stage were dominated by sea scorpions up to 7 feet (2 m) long.

In the Ordovician period, jawless fish were followed by sharks and bony fish.

In the Cambrian period, sponges, segmented worms, and the first hard-shelled animals came on the scene.

In this period, the first multi-celled animals evolved. They had soft bodies and resembled present-day worms and jellyfish.

The earliest forms of life on Earth were algae and single-celled bacteria that appeared in the seas.

Origin of the Earth

LIFE EVOLVED *from a few primitive marine organisms to a wide variety of marine and land animals in the 4,000 million years before dinosaurs appeared. (Right) A trilobite.*

simultaneously. This allowed them to travel much greater distances without stopping.

The oldest known dinosaur fossils were uncovered from 228-million-year-old rocks from Argentina and Mada-gascar. The fossils from Madagascar represent primitive prosauropods, and those from Argentina are closely related to early theropods such as *Herrerasaurus* and *Eoraptor*.

THE RULE OF THE DINOSAURS

For 150 million years, the greater part of the Mesozoic era, dinosaurs were the dominant land animals on Earth.

For the first third of the Triassic, archosaurs were only minor components of the fauna. They were overshadowed by the synapsids, close relatives of mammals. By the end of the Triassic, the synapsids were small creatures overshadowed by the archosaurs—especially the dinosaurs. The dinosaurs continued as the dominant land vertebrates for the remainder of the Mesozoic.

TRIASSIC DINOSAURS

The earliest dinosaurs were bipedal predators, much like the early theropods *Eoraptor* and *Herrerasaurus*. Most of the known Triassic theropods were outwardly similar, the largest being approximately 6 feet 6 inches (2 m) long. The dominant theropod lineage of the time, known as the coelophysids, included the *Coelophysis*, which was found in New Mexico.

Plant-eaters appeared early. Both early prosauropods and sauropods are known from the start of dinosaur history: one of the best known Triassic dinosaurs is the

CAMPTOSAURUS *was a bird-hipped herbivore of the late Jurassic. Remains of this dinosaur, which grew to 20 feet (6 m) long, have been found in Europe and North America.*

prosauropod *Plateosaurus* from Germany, and the oldest sauropod is *Antetonitrus* from South Africa. We have evidence for exclusively plant-eating ornithischians, such as *Lesothosaurus* from southern Africa.

Although later sauropodomorphs and ornithischians were diverse, and differed greatly from the theropods, the earliest herbivores still retained some features of the earliest dinosaurs, in that they were relatively small and could move on two feet.

INCREASING DIVERSITY

During the Triassic and early Jurassic, when there was only one major landmass and the

The astonishing claim that birds are dinosaurs cannot be ignored by fellow palaeontologists.

Sunday Times, *March 17, 1974*

climates were relatively uniform, dinosaurs moved quite easily from one region to another. As a result, dinosaur finds from the late Triassic and early Jurassic generally tend to represent very similar groupings, wherever they are from. They consist typically of one or a few small theropods; a prosauropod, which is usually the largest dinosaur of the group; and a small, bipedal ornithischian.

As the Jurassic progressed, theropods continued to diversify. Most of those that have been discovered were relatively large animals, but a few smaller theropods were also found.

We know, however, that there must have been a wide diversity of small theropods, because the first bird, *Archaeopteryx*, is from the late Jurassic. Most of the small theropods that were closest to birds are known only from the Cretaceous. If they shared a common ancestor with

WHY SO LONG?

We will never know for certain why dinosaurs enjoyed such a long dominance. Some argue that their advanced sytem of locomotion, which allowed them to breathe while walking or running, gave them an edge over other Triassic groups, including the synapsids, which did not evolve a completely upright posture until much later. Others contend that dinosaurs were simply filling an ecological void caused by the extinction of the dominant tetrapods of the early Triassic. Neither group, they maintain, was "better" than the other. Dinosaurs simply managed to survive whatever catastrophe befell the dominant early and middle Triassic vertebrates and to radiate before the other groups could.

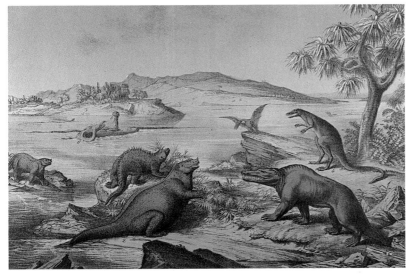

birds, then these groups must also have been present in the Jurassic, even though their fossils remain undiscovered.

CHANGING HABITS

The largest herbivorous vertebrates of all time, the sauropods, were the most common plant-eaters of the late Jurassic and early Cretaceous. They seem to have been most diverse during the late Jurassic, but the largest known forms lived during the early Cretaceous.

Although sauropods still existed at the end of the Cretaceous in some parts of the world, they were less common in the middle and late Cretaceous.

Ornithischians diversified broadly during the Jurassic. During the late Jurassic, certain groups of armored thyreophorans—including the stegosaurs—were common throughout North America, Eurasia, and Africa. Primitive ornithopods also appear during the Jurassic, but they were not major components of the fauna.

During the Cretaceous, the structure of herbivorous dinosaur communities under-went a radical shift. Sauropods, for example, often became minor components and the ornithopods became dominant ones. Late in the Cretaceous, ankylosaurs and, in some locations, ceratopsians, were very common.

REGIONAL VARIATIONS

As the continents diverged throughout the Mesozoic, dinosaur faunas around the world became increasingly distinctive—North American groupings, for instance, no longer looked like South American or Asian assemblages. The sauropods, for example, which were common throughout all parts of the world during the Jurassic, had become common only in Madagascar, South America, and India by the late Cretaceous.

Some groups do not appear ever to have had a worldwide distribution. The ceratopsians and pachycephalosaurs, for example, are known only from western North America and from eastern Asia. These regions, which were connected to one another during the late Cretaceous, were, however, largely isolated from the other parts of the world.

MORE SPECIES *of dinosaurs evolved during the Cretaceous period than during the Triassic and Jurassic combined. This group shows how diverse these animals had become by the Cretaceous. They are, from left,* Triceratops, Corythosaurus, Pachycephalosaurus, Saltasaurus, Euoplocephalus, *and* Tyrannosaurus.

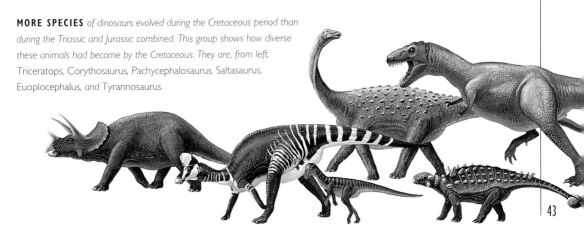

THE PTEROSAURS

Those animals that have the ability to glide or fly have some distinct advantages over their ground-bound relatives.

Lots of organisms travel through the air. Some—including many plants and a few invertebrates, such as spiders—use the wind to spread their seeds or young. The ability to actively move through the air—to fly or glide—has evolved only a few times, but many of the groups that have developed this skill have become hugely diverse. Flying or gliding allows broad dispersal and opens niches not available to non-flyers. The insects, of course, are the most successful such group, but several gliding and flying groups have appeared among the vertebrates.

AIRBORNE VERTEBRATES

The first true vertebrate flyers—animals that are capable of generating lift with their wings—were the pterosaurs. They first appeared in the late Triassic and they survived right to the end of the Cretaceous. Pterosaurs were archosaurs. Although they were probably close relatives of dinosaurs, they were not, as is often popularly supposed, real dinosaurs.

Pterosaurs are often called pterodactyls because of the structure of their wing. "Pterodactyl" means "wing finger," and the wing did, indeed, consist largely of a single finger. Pterosaur arms had four fingers. The first three were small, but the fourth was of an enormously length. A membrane of skin extended from the tip of this finger to either the side of the body or the hindlimb. A few pterosaur fossils preserve the imprint of the wing membrane and show that it was not just a thin flap of skin; rather, it was given support by slender collagen rods that would have kept it relatively rigid.

Pterosaur wings were thus similar to bat wings in that the flight surface was skin supported by the hand, but whereas pterosaurs relied on a single finger from the hand, bats use three or four. An additional skin membrane stretched between the animals' hindlimbs.

LARGE AND SMALL

The earliest pterosaurs were crow-sized animals. They had jaws filled with sharp teeth and probably fed on a variety of small animals, including vertebrates and insects. They had long, slender tails. More advanced pterosaurs differed from their Triassic ancestors in several ways—the tail was greatly shortened, the teeth were lost, and toward the end of the Cretaceous, these animals became very large. In fact, some of the last pterosaurs were the largest known animals ever to take to the air and fly. *Quetzalcoatlus*, from the late Cretaceous of North America, may have had a wingspan of up to 40 feet (12 m)—greater than the wingspan of some types of fighter aircraft!

Throughout the Mesozoic, the pterosaurs filled a large number of roles. Some retained the small size and toothy mouth that they had inherited from their ancestors, but others modified their anatomy to acccommodate different needs. A number of them, including *Pteranodon* from the Cretaceous of North America, which had a wingspan of 23 feet (7 m), may have lived like modern gulls or albatrosses, soaring over the ocean in search of fishes. Other pterosaurs had heads that were shaped like those of sandpipers or flamingoes.

One pterosaur fossil of the Triassic—*Sordes pilosus*, from Kirghizia in eastern Europe—is preserved with the remains of short, hairlike fibers around the body. Some believe this was an external covering, similar to down. This is why some recently reconstructed pterosaurs have a "fuzzy" look about them.

A PTEROSAUR *of the late Jurassic, Rhamphorhynchus, with its distinctive long tail, grew about as big as a modern-day seagull.*

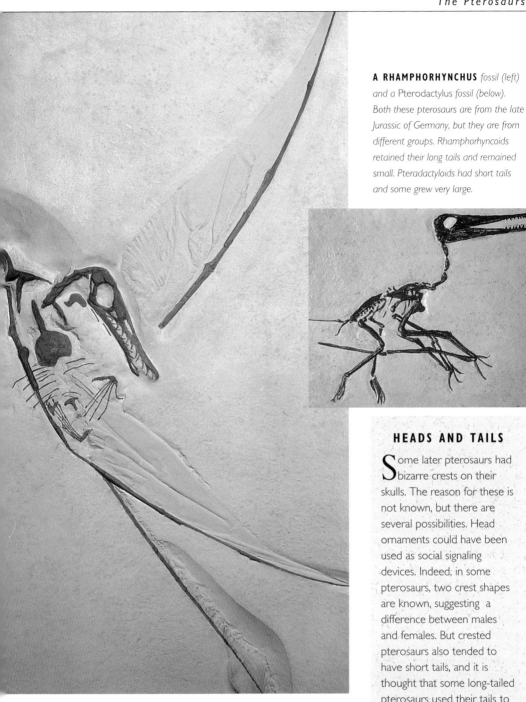

A RHAMPHORHYNCHUS *fossil (left) and a Pterodactylus fossil (below). Both these pterosaurs are from the late Jurassic of Germany, but they are from different groups. Rhamphorhyncoids retained their long tails and remained small. Pteradactyloids had short tails and some grew very large.*

HEADS AND TAILS

Some later pterosaurs had bizarre crests on their skulls. The reason for these is not known, but there are several possibilities. Head ornaments could have been used as social signaling devices. Indeed, in some pterosaurs, two crest shapes are known, suggesting a difference between males and females. But crested pterosaurs also tended to have short tails, and it is thought that some long-tailed pterosaurs used their tails to help steer while flying. In fact, the late Jurassic pterosaur *Rhamphorhynchus* had a diamond-shaped keel at the tip of its tail. Some experts think crests may have helped short-tailed pterosaurs to steer. A flying model of *Quetzalcoatlus* was built during the 1980s, and engineers steered the model by turning the head, relying on the crest to control the direction of flight.

MOVING AROUND

There is some disagreement about how pterosaurs moved on the ground. Most researchers believe they were quadrupeds; some maintain they were bipedal. The four-footed view is supported by the structure of the hindlimb and wing membrane, as well as by footprints believed to have been formed by walking pterosaurs. They show what may be the impressions of both feet and folded wings.

We do not know how pterosaurs evolved flight. The first known pterosaurs in the Triassic already had a complete flight apparatus. We do not have the range of transitional forms that connect birds with their dinosaurian precursors. Most researchers assume that the precursors of pterosaurs were tree-dwelling archosaurs that jumped between trees. However, fossil evidence for this is scarce and inconclusive.

MARINE REPTILES

The return to a life in water has been a recurring theme for many different groups of reptiles.

No sooner had reptiles adapted to a life out of water than they started returning to a life in the seas. Time spent foraging in the productive shallows gradually led them to become increasingly proficient in the water. Structures needed for life on land—such as efficient respiratory and water-control systems—were no doubt an asset in this new environment.

ICHTHYOSAURS AND SAUROPTERYGIANS

The ichthyosaurs and sauropterygians appeared in the Triassic, and their careers as dominant marine predators lasted at least 150 million years. The earliest members of these groups were already so adapted to an aquatic life that the identity of their terrestrial ancestors has yet to be established. The ichthyosaur tail became shaped like that of a fast-swimming fish, the head merged with the body as it does in dolphins, and the teeth and jaws were well adapted to a diet of fish and cephalopods.

Many groups of sauropterygians became very successful marine predators. The strange

THE GREAT WHITE SHARK *is one of the most powerful, and feared, predators in today's oceans. It is a descendant of the ancient giant lamnid sharks that first appeared in the world's oceans during the Cretaceous.*

placodonts acquired heavy bones and strong, crushing teeth, and seem to have fed upon shellfish. Nothosaurs evolved a flattened skull with powerful jaw muscles on the end of a long, flexible neck which could be swiped rapidly sideways through the water to catch agile fish. Nothosaurs used their forelimbs as well as—or even instead of—their tail for propelling themselves through the water.

The descendants of the nothosaurs—the plesiosaurs—made their appearance in the very early Jurassic. In plesiosaurs all four limbs became paddles which were used for swimming

OPHTHALMOSAURUS, *a late Jurassic ichthyosaur, was remarkable for its huge eyes. It probably fed at night.*

through the water, and the tail was greatly reduced.

The two major groups of plesiosaurs modified the nothosaur feeding system to opposite extremes. The necks of long-necked plesiosaurs grew to almost ridiculous proportions—the elasmosaurs of the Cretaceous had necks with more than 80 vertebrae. The head was small, and could be moved quickly when catching prey.

In contrast, the pliosaurs strengthened the large skull of their nothosaur ancestors so it could handle large prey—and because these large prey were attacked head on, the flexible neck was shortened and strengthened to stabilize the massive skull. Genera such as *Liopleurodon* and *Kronosaurus* were some of the largest predators of all time.

FOSSIL SKELETONS
of Cretaceous marine reptiles such as pliosaurs (right) and ichthyosaurs, like Platypterygius (below), have been found in most parts of the world. This marine creature grew 23 feet (7 m) long. In the ichthyosur fossil below, a newborn young can be discerned just near the base of the mother's tail.

Turtles, Crocodiles, and Lizards

Turtles have made two major incursions to the sea. One was in the Jurassic, while the other started in the Cretaceous and has lasted to modern times. Most turtles have been medium-sized animals, feeding on bottom-dwelling animals and plants. During the Cretaceous, however, a number of species became very large. All turtles lay eggs; the hatchlings spend their early years far out to sea, moving back to the adult feeding grounds between 5 and 20 years later. This life cycle may explain why turtles survived the mass extinction at the end of the Cretaceous. Even if all the adults had died, there would have been at least five years' worth of young animals waiting to recolonize the habitat. Crocodiles, too, started

a return to the sea in the Jurassic. The teleosaurs were highly adapted for a life at sea: They appear to have preyed upon fish, squid, and even other reptiles. They adopted a form of the underwater flight also employed by plesiosaurs and turtles, where the hindlimbs were used as underwater wings.

Lizards are successful today as small land-living animals, but they, too, have produced several marine lineages. The varanids (goannas and monitors) are the largest modern lizards—yet in the late Cretaceous a group of them returned to the sea to become the spectacular mosasaurs. Even today there are several species of varanid, including the water monitor and the

Komodo dragon, that are quite at home in the ocean. The marine iguana of the Galapagos is perhaps the best example of a lizard returning to a life in the sea.

Sharks

Sharks are not reptiles, but their history has been an important part of the marine reptile story. When reptiles first started to become marine predators, the large sharks were sluggish and not very good swimmers. By the Cretaceous, however, new groups of sharks—the carcharinids (reef and whaler sharks) and the lamnids (white and mako sharks)—had appeared. The lamnids are warm-blooded and are fast, powerful hunters. When the large pliosaurs died out in the middle of the Cretaceous, their place was taken by huge lamnids, even more powerful than the great white shark of today.

KRONOSAURUS, *which is known from Queensland, Australia, was a pliosaur of the early Cretaceous. It grew more than 40 feet (12.5 m) long, twice the length of a modern great white shark.*

OTHER LAND VERTEBRATES

Throughout the Mesozoic era, many interesting groups of animals, including some that are still alive today, existed alongside the dinosaurs.

For most of the Triassic, dinosaurs—and, more generally, archosaurs—were a minor part of the land fauna. During the early Triassic, synapsids were the most diverse group of large-bodied vertebrates on land.

MAMMAL BEGINNINGS

The class Synapsida now includes mammals, but during the early Triassic it comprised a large number of bizarre creatures that we used to call "mammal-like reptiles." Today scientists no longer include the early synapsids in the class Reptilia. However, when they first appeared during the Paleozoic era, they did look rather reptile-like,

especially as their limbs sprawled out to the side. The synapsids of the early Triassic were more sophisticated. In them, advanced features that we see in living mammals—such as specialized teeth—were blended with more primitive characteristics found in ancestral amniotes—such as jaws with more than one bone and a less advanced way of moving.

As the Triassic progressed, synapsids became smaller and more mammal-like. Synapsids of the latest part of the Triassic looked a great deal like modern shrews. Their cheek teeth had multiple cusps, their ears had multiple bones, and each side of their

jaws had only one bone. As well, they probably had hair. True mammals did not appear until the Jurassic, but these late Triassic synapsids probably lived much like small mammals.

Synapsids were present throughout the Mesozoic, but from the late Triassic through to the end of the Cretaceous, they remained small. The largest of them was no bigger than a modern domestic cat. The first true mammals—members of the group that included living monotremes, marsupials, and placentals—appeared during the Jurassic, and although small, they were probably diverse. Nevertheless, they were not the predominant land vertebrates they would become during the Cenozoic era.

SNAKES *first appeared during the Cretaceous, and the oldest known snake, which is 80 million years old, is known from Cretaceous rocks in South America. The fossil below is an undescribed Tertiary species from Messel in Germany.*

THE JAWS *and teeth of modern crocodiles (left) bear a striking resemblance to those of synapsids, such as* Dimetrodon *(below).* Dimetrodon, *however, was not a reptile but a distant relative of today's mammals.*

REPTILES

Other groups that were common in the Triassic have no close surviving relatives. Included among them were the rhynchosaurs. These were bizarre parrot-beaked, plant-eating reptiles, which were distantly related to the archosaurs. They are common in some late Triassic deposits. However, there is no fossil evidence to indicate that they survived beyond the Triassic. Other lizard-like reptiles, such as the procolophonids, may have been distant relatives of modern-day turtles.

Members of the order Crocodyliformes were very common and diverse during the Mesozoic. Many people think of these animals as "living fossils," because fossil crocodyliforms from the Mesozoic often outwardly resemble modern crocodilians—the alligators, crocodiles, and gharials. It is true that many Mesozoic crocodyliforms appear to have been "crocodile-shaped," but it would be incorrect to suggest that they have remained unchanged to the present day.

The earliest crocodyliforms were probably exclusively land-dwellers, and they had longer limbs and deeper skulls than their present-day counterparts. They may have ambushed their prey, much as the earliest archosaurs did. Throughout the Jurassic and Cretaceous, there was a variety of crocodyliforms. They included several lineages that adapted to life in the open sea, complete with limbs modified as flippers and, in a few groups, a tail that bent downward to accommodate a tail fin, much like that of an ichthyosaur. The first true crocodilians appeared during the late Cretaceous and, until the Tertiary, they lived alongside several other crocodyliform groups.

Other reptilian groups were common during the Mesozoic. Sphenodontia today includes only one species—the tuatara (*Sphenodon punctatus*) of New Zealand. During the Triassic and Jurassic, however, sphenodonts were very diverse and

DIMETRODON *was a synapsid—a mammal relative—that lived in the Permian period.*

included some semiaquatic groups. Although most non-dinosaurian reptiles of the time looked like lizards, true lizards did not appear until the Jurassic.

Lizards were very diverse during the Cretaceous and included some large terrestrial predators and one of the most important groups of marine reptiles, the mosasaurs. A particular group of legless lizards—the snakes—appeared during the Cretaceous.

There is fossil evidence that turtles, including some aquatic turtles, existed in the late Triassic and the Jurassic. This is not surprising: Turtle shells are extremely durable, which makes them excellent candidates for fossilization. Modern turtles are remarkably similar to their ancient counterparts. However, true sea turtles—the ancestors of today's sea turtles—did not appear until the Cretaceous.

DEATH OF THE DINOSAURS

*Theories abound, but mystery still surrounds the extinction
of the dinosaurs at the end of the Cretaceous period.*

Dinosaurs dominated the landscape for most of the Mesozoic, and, in a sense, they are still diverse today—they live on in the birds, which are their direct descendants. However, the animals popularly thought of as dinosaurs ceased to exist at the end of the Cretaceous and within a few million years, large mammals and birds were filling the roles once played by giant nonavian theropods and duckbills. Why did these animals disappear?

Just what caused the extinction of the nonavian dinosaurs is still hotly debated. Today there are very few places on Earth where the Cretaceous–Tertiary—often called the K/T—boundary is preserved in a sequence of land or freshwater sediments. We have a good idea about what happened in the marine realm, but only a few dinosaurs—in the form of aquatic birds—lived in the sea. So we are forced to generalize from what the few Cretaceous–Tertiary sites reveal and apply it to the whole world.

AN ANCIENT PUZZLE

Even events of recent times, observed by large numbers of people and recorded on videotape, are subjects that evoke heated controversy. The assassination of the United States president John F. Kennedy, which occurred less than half a century ago, is a case in point. If we cannot know for certain what happened then, what chances have we of solving a mystery that is millions of years old, with only the evidence of what is probably a very incomplete fossil record?

As far as we can tell, up to 70 percent of species of marine organisms became extinct at the end of the Cretaceous. Marine invertebrates were particularly affected. Prominent groups, such as ammonites—relatives of today's nautilus and squids—disappeared entirely. The extinctions also claimed many species of marine reptiles: Mosasaurs and plesiosaurs also disappeared.

SUDDEN OR GRADUAL?

We know less about what happened on land, but it is clear that all the nonavian dinosaurs disappeared at the end of the Cretaceous. So too did the pterosaurs and many groups of smaller vertebrates.

Some researchers have argued that the disappearance

AN IMPACT *that occurred 300,000 years ago resulted in the 3,000-feet (900-m) wide Wolfe Creek Crater in Western Australia's Wolfe Creek Crater National Park.*

A DRAMATIC EXAMPLE *of the Cretaceous–Tertiary boundary (above) can be seen in this hillside north of Saltillo, Mexico. This meteorite (right), which is believed to have come from Mars, is 4,500 million years old.*

of nonavian dinosaurs occurred very suddenly, and that several lineages died out within the space of a few thousand years—a relatively brief time in terms of the Earth's history. Others, however, contend that the extinction happened more gradually, and that some groups disappeared well before

the Cretaceous–Tertiary boundary. The evidence we have, which is restricted to only a few places, is not sufficient to resolve this question. While it seems likely that the marine invertebrates disappeared rather suddenly, the extinction of land animals may have been a more gradual process.

POSSIBILITIES

Many theories have been proposed to explain the extinction of the nonavian dinosaurs. These include poisoning from the first flowers, the destruction of eggs by early mammals, radiation poisoning from a distant supernova, and widespread plagues. All of these explanations, while they may seem plausible, leave many important questions unanswered. Flowering plants, for example, had evolved tens of millions years before dinosaurs became extinct. Mammals, too, had shared the land with dinosaurs for well over 100 million years. Most seriously, however, these theories fail to come to terms with the mass extinctions that occurred at the end of the Cretaceous—why so many groups of organisms, apart from dinosaurs, suffered at this time.

THE HOBA METEORITE, *from Namibia, is the largest meteorite yet found. It is between 410 and 190 million years old.*

CURRENT THEORIES

Most present-day paleontologists subscribe to one or more of three major theories to explain the Cretaceous-Tertiary mass extinction. All three have good evidence to support them, and all three may have been important factors, reinforcing each other in a vicious cycle of environmental devastation.

The first involves climatic changes that were already under way during the Cretaceous. Global temperatures fell at the end of the Cretaceous, and seasonal variations became more pronounced. Differences in summer and winter temperatures, especially at high latitudes, increased markedly. This phenomenon may have been partly the result of major changes in sea levels, which would certainly have increased the amount of habitat that was available for marine organisms.

Rising sea level changes, too, may well have resulted in increased fragmentation of terrestrial habitats. Stretches of forests that were formerly continuous, for example, may have become broken up. This in turn may have fragmented

PROFESSOR ROBERT ROCCHIA, *in his laboratory in France in 1993, holds 65-million-year-old samples of Cretaceous chalk and Tertiary limestone from the Cretaceous–Tertiary boundary.*

large concentrations of animals, resulting in smaller populations, which would be more vulnerable to environmental changes.

However, while climate and changes in sea levels could clearly affect the survival or extinction of living species, they would seem unlikely to bring about a sudden demise. If, as many researchers maintain, dinosaur extinction was sudden, other factors must have played a part.

The second major theory involves extensive volcanic eruptions. Several parts of the world at the end of the Cretaceous were experiencing a huge amount of volcanic activity. The most spectacular evidence for this is the Deccan Traps, a massive set of flood basalts in India.

THIS SEM *(Scanning Electron Micrograph) of magnetite in rock lends support to the theory of asterod impact at the end of the Cretaceous. (Right) A volcanic eruption in the Galapagos Islands.*

The third, and most dramatic, theory postulates the impact of a comet or asteroid at the very end of the Cretaceous. That an impact occurred is supported by several lines of evidence. Most convincing is the existence of certain elements, such as iridium, that are enriched in sediments laid down at the Cretaceous-Tertiary boundary. These elements are rare in the Earth but are common in extraterrestrial materials. Particles of shocked quartz—minerals that have been subjected to intense pressure—are also found in the boundary sediments. Geologists now believe we have found the crater caused by this impact in the Yucatàn Peninsula in Mexico, although it is buried deep underground.

Many of the environmental effects caused by volcanic activity and asteroid impact would be very similar. Both cause large amounts of dust and ash to fill the upper atmosphere, preventing sunlight from reaching the surface and impacts on the survival of plants. Any significant loss of plant life would certainly have had a cascading effect on other organisms. Plant-eating animals would obviously have suffered, as would carnivorous animals that preyed on them.

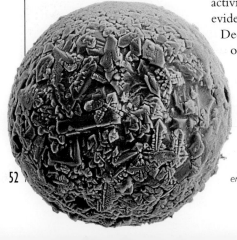

There is evidence for a temporary disturbance in plant populations in North America at the Cretaceous-Tertiary boundary, because at that point we find a sudden increase in fern spores. Ferns tend to become very common if other plants, such as angiosperms and conifers, are killed off. This suggests that something caused large portions of North America—near the impact site in the Yucatàn Peninsula—to be stripped, quite suddenly, of most vegetation. Flowering plants and conifers came back in the Tertiary, but the large plant-eating dinosaurs were no longer there to eat them.

The solution to the mystery of the mass extinction may continue to elude us. It is quite possible that all three factors cited in currently favored theories played an important part. There are certainly good reasons to

Our little systems

have their day;

They have their day and

cease to be.

In Memoriam,
ALFRED, LORD TENNYSON
(1809–1892), English poet

suggest there is a link between climate changes, volcanic activity, an asteroid impact, and mass extinctions, but the solid evidence that would clinch the argument does not yet exist. One thing we can agree on is that the early Tertiary world was very different from the late Cretaceous one. The only dinosaurs that seemed to have survived into the Tertiary were the feathered ones that flew across the Cretaceous-Tertiary boundary.

GLOBAL WARMING

As well as blocking the rays of the sun, the vast clouds of dust and ash that result from both an asteroid impact and volcanic eruptions can also trap ultraviolet radiation within the atmosphere, causing global warming. Volcanic eruptions usually include emissions of greenhouse gases, such as carbon dioxide, which can further enhance this warming. Even if it lasted only for a relatively short time, a sudden increase in temperature could in turn have contributed to extinctions by creating an environment that was unsuitable for many animals. Even more dramatic effects of an asteroid impact—such as giant tidal waves, global wildfires, and acid rain—have also been proposed, but there is no conclusive evidence that these occurred.

THE SURVIVORS OF MASS EXTINCTION

Large numbers of animal species survived the cataclysmic events at the end of the Cretaceous.

THIS SMALL MAMMAL, Purgatorius, *survived the cataclysmic events that occurred at the end of the Cretaceous.*

The mass extinction that occurred at the end of the Cretaceous destroyed over half the Earth's animals. The most familiar victims of the extinction were the nonavian dinosaurs and several important groups of marine invertebrates, such as the ammonites. Often ignored, however, are the important groups of animals that survived. They raise interesting questions about exactly what happened at the boundary of the Cretaceous and the Tertiary.

AQUATIC ANIMALS

The group that seems to have suffered least from changes at the Cretaceous–Tertiary boundary was the bony fishes. In both marine and freshwater settings, these animals appear to have maintained their former diversity.

There is little evidence to suggest that some other important groups that were dependent on fresh water— such as amphibians, freshwater turtles, and crocodylians— became any less diverse than before. This is curious, because scientists have long assumed that acid rain was one of the most harmful consequences of the asteroid impact that occurred about that time. Researchers believe that materials ejected from the crust by this impact would have interacted with atmospheric water to form acids that fell to earth as rain. But, strangely, the groups that have suffered most from the effects of acid rain in modern times—the freshwater fishes and amphibians—fared better than most in the events at the boundary of the Cretaceous and the Tertiary.

MAMMALS AND BIRDS

We cannot be certain about how badly mammals suffered at the Cretaceous–Tertiary boundary—or, indeed, whether they were affected at all. We do know that mammal species did increase in diversity early in the Tertiary. It also appears that some mammal groups became extinct at the end of the Cretaceous. Some researchers, however, argue against this. They maintain that, rather than becoming extinct, these groups simply evolved into the forms for which we have fossil evidence in the Tertiary.

One factor that may have saved many mammals was their size. Cretaceous mammals were small and may well have avoided the effects of climatic, volcanic, and other changes by burrowing into the ground and hiding.

Among flying vertebrates, pterosaurs died out completely, but dinosaurs left their mark in the form of their direct descendants, the birds, and in

AS YET *we do not know for certain why many animals became extinct, while others survived the Cretaceous–Tertiary boundary.*

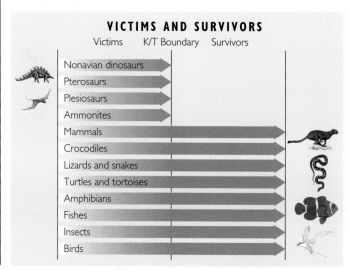

VICTIMS AND SURVIVORS

Victims	K/T Boundary	Survivors

Nonavian dinosaurs
Pterosaurs
Plesiosaurs
Ammonites
Mammals
Crocodiles
Lizards and snakes
Turtles and tortoises
Amphibians
Fishes
Insects
Birds

LANDSCAPES *such as Death Valley, California (above), characterize much of the post-Cretaceous world. Mammals that survived were smaller than many modern mammals (right).*

their more distant relatives, the crocodiles and alligators. However, not all avian groups outlived the Cretaceous. One very diverse group, the enantiornithines, which were the dominant birds of the Cretaceous and existed on all the continents, does not seem to have survived into the Tertiary.

THE FOSSIL RECORD

In recent times, biologists have been debating whether the fossil record accurately reflects the extent to which the diversity of birds and mammals changed across the Cretaceous–Tertiary boundary. According to the fossil record, these groups, already diverse in the Cretaceous, later became even more diverse, with large

numbers of mammal and bird lineages appearing early in the Tertiary. This suggests that these animals adapted to the new conditions and diversified, with mammals, especially, filling the roles played by nonavian dinosaurs in the Mesozoic.

Some molecular biologists question this pattern. If we assume that changes in genes occur at a roughly constant rate over time, we can estimate the time at which living groups diverged from each other by counting the number of differences in their genes. Using this approach, some biologists have calculated that divergences

between living mammal and bird groups occurred deep in the Cretaceous—long before these groups appear in the fossil record. This suggests that, while large numbers of bird and mammal lineages were present during the Cretaceous, there is no trace of them in the fossil record.

It is perfectly possible that the fossil record is incomplete. It is equally possible, though, that there are problems with the way biologists have been calibrating the "molecular clocks" that they use to help them calculate the time that divergences occurred. Further research may eventually reveal the truth.

TODAY'S DINOSAURS

The evolutionary link between present-day birds and long-extinct dinosaurs was one of the most exciting scientific discoveries of recent times.

Birds are living descendants of the dinosaurs that disappeared at the end of the Mesozoic era, 65 million years ago. Look at footprints left behind by a bird, and you will usually see the imprints of three toes spreading out from the rest of the foot. Now look at the footprints left by Mesozoic theropod dinosaurs, and you will see a similar pattern. Is there a reason for this?

A DINOSAUR ANCESTRY
The idea that birds are living dinosaurs is not a new one. Some of the most highly regarded anatomists of the late 1800s thought birds were the descendants of dinosaurs but, for most of the 20th century, scientists thought they were descended from some other kind of archosaur. It was not until the 1970s that birds began to be restored to their dinosaurian perch. Nowadays, a dinosaurian ancestry for

birds is just as well supported as a mammalian ancestry is for humans.

Perhaps the most famous fossil of all time is that of *Archaeopteryx*, the earliest known bird. Seven skeletons have been found, all from the late Jurassic lithographic limestones of Solnhofen, in southern Germany.

Archaeopteryx preserves a mixture of features that made it a "missing link" between birds and other reptiles. The jaws have teeth and the tail is long and bony—both reptilian features—but at least in some of the specimens, there are clear impressions of feathers on the body. Some specimens of *Archaeopteryx* were at first misidentified as the small theropod *Compsognathus*, because the impressions of feathers were very faint.

The first *Archaeopteryx* skeleton was found in 1861, only two years after the publication of Darwin's *On the Origin of Species*, which gave the scientific community powerful evidence that species changed over

time. And it was a critical discovery for those studying the origin of birds—modern birds are so modified for flight that it is difficult to link them to any earthbound group of animals. *Archaeopteryx* was a primitive enough bird to still retain many features of the nonavian relatives of birds.

DISCOVERY AT DINNER
The prominent late 19th-century scholars who believed in birds' dinosaurian ancestry based their conclusion on skeletal similarities between birds and dinosaurs, especially in the hindlimb. Birds and dinosaurs both have a hole in the hip socket, and the ankle and foot of a chick look much the same as those of a theropod dinosaur. According to legend, Thomas Henry Huxley (1825–1895) first made the bird–dinosaur connection at a formal dinner after a day spent in the museum examining fossil dinosaur bones. The main course was poultry and, as Huxley ate, he noticed on his plate features that he had seen earlier in the museum.

Early in the 20th century, the idea that birds were derived from dinosaurs fell out of favor. Anatomists pointed out that, despite their shared features, birds and theropods differed in one crucial respect: Birds have a very large set of collarbones,

IN LATE 1996 *a new small theropod dinosaur, which was described as Sinosauropteryx, meaning "Chinese lizard wing," was discovered in China. It had hairy fibers covering much of its body—from head to tail, along the sides, and along the arms and legs. It is not yet clear whether these fibers were feathers.*

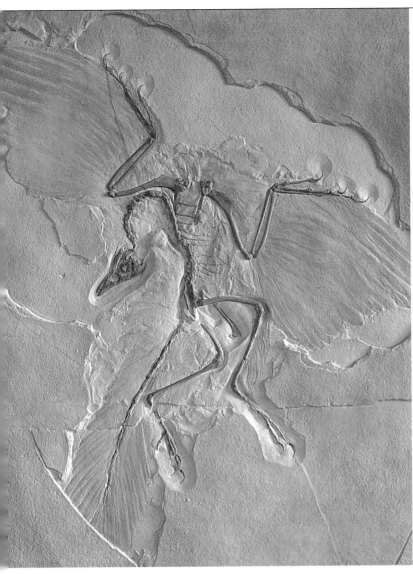

THE ILLUSTRATIONS *below show stages in the evolution of birds from dinosaurs. In the course of this evolution, the jaws and teeth gave way to a toothless beak, the bony tail was lost, and the nature of the hips changed. The four-toed hind foot, though modified, was retained.*
❶ Compsognathus
❷ Archaeopteryx
❸ *A living bird*

or clavicles, that are fused together to form the wishbone. No dinosaurs then known had a collarbone. Scientists at that stage believed that once a structure is lost during the process of evolution, it cannot be regained. Working from this premise, they argued that dinosaurs could not have given rise to birds. All the features they shared must have evolved convergently. The detailed similarities in the hindlimb and hip were viewed simply as being adaptations to the task of standing on two legs. The first dinosaurs that had clavicles were discovered in the 1920s, but the discovery largely went unnoticed for many decades.

A THEORY REAPPRAISED

The turning point came in 1964, when John Ostrom of Yale University discovered the remains of a small theropod, *Deinonychus*, in lower Cretaceous rocks of Montana. Ostrom was struck by a number of curious similarities between *Deinonychus*—and other small theropods—and primitive birds. The construction of the dinosaur's hands and birds' feet was very similar. *Deinonychus* had a flattened half-moon-shaped bone in the wrist that limited movement of the hand, much like a similar element of the bird wrist. The bones of the pelvis were also very similar in

shape. *Deinonychus*, then, had numerous striking features in common with birds. Based on his work with *Deinonychus*, Ostrom in the 1970s resurrected the hypothesis that birds were derived from theropod dinosaurs.

A NEW SYSTEM OF GROUPING

Beginning in the 1980s, modern cladistic analysis—a system of classification developed in the 1950s by the German biologist Willi Hennig, in which all organisms that have a common ancestor are grouped together in a "clade"—was applied to fossil vertebrates. The first "cladistic" study to include birds and dinosaurs was published by Jacques Gauthier, now of Yale University, in 1986. Gauthier's analysis confirmed Ostrom's belief that small theropods such as *Deinonychus* were the closest extinct relatives of birds. To date, all such cladistic analyses strongly support a close relationship between birds and theropod dinosaurs.

Like birds, all dinosaurs have an inturned femoral head and open hip socket. Again like birds, all theropods have thin-walled hollow limb bones, complex air sacs in the skull and vertebral column, a three-toed foot, and a hand dominated by the thumb, index finger, and middle finger. As we look at smaller groups of theropods, we see the forelimb and hand become increasingly

birdlike, with the complete loss of the fourth and fifth fingers and the development of a specialized wrist. The tail becomes stiffened and reduced in length, and in those theropods closest to birds, including *Deinonychus*, the pubis points down or back, not forward. As well, we now know that many groups of theropods not only have collarbones but also true wishbones.

DO ONLY BIRDS HAVE FEATHERS?

Until quite recent times, feathers were considered to be exclusive to birds, including *Archaeopteryx*. During the course of the 1990s, however, several new discoveries from north-eastern China have proved otherwise. A fossil locality in Liaoning province began to yield the remains of small theropod dinosaurs, some of them with curious fibrous structures surrounding the body.

The first of these dinosaurs to be announced was *Sinosauropteryx*. This was an animal about the size of a turkey that was very similar to *Compsognathus*, and it preserved a halo of short hairlike structures, which, it is now thought, may be the precursors of true feathers. In some of the theropods discovered at Liaoning, including *Beipiaosaurus* and *Sinornithosaurus*, short fibers, much like those of *Sinosauropteryx*, have been preserved, but in two others, *Protarchaeopteryx* and *Caudipteryx*, the feathers are unambiguous—they have a central shaft (rachis) as well as fibers (barbs).

As a result of these discoveries, we can no longer regard feathers as belonging exclusively to birds. *Sinosauropteryx* and *Compsognathus* are relatively primitive coelurosaurian theropods, which means that feathers, or their precursors, were present on many groups of theropods, even though these are not often preserved. We can no longer simply draw a line that separates bird and theropod; modern birds are clearly living members of the Dinosauria, just as humans are living members of the Mammalia.

JOHN OSTROM

John Ostrom was one of the most distinguished paleontologists. He was a professor and curator of vertebrate paleontology at Yale University's Peabody Museum of Natural History for many years, and he was one of the driving forces behind the "dinosaur renaissance" of the early 1970s, in which the old image of dinosaurs as overgrown reptiles was replaced by an image of more dynamic, energetic animals. Ostrom published many landmark papers during his career, including descriptions of the small theropod *Compsognathus* and important analyses of the origin of flight, based largely on first-hand observations from the primitive bird *Archaeopteryx*. His description of *Deinonychus* stands as a turning point in the history of vertebrate paleontology.

DEINONYCHUS *was a ferocious predator. In this painting (left) two members of the species spar in foreground while others in the pack attack a* Tenontosaurus. *(Below) A fossil skeleton of* Sinosauropteryx *shows the short feathers that covered the body.*

THE ORIGINS OF FLIGHT

The knowledge that birds are dinosaurs has interesting implications for our understanding of the origins of flight. Paleontologists used to believe that all flying or gliding vertebrates—bats, birds, pterosaurs, sugar gliders, and the like—evolved from tree-dwelling animals that jumped from one perch to the next. As some of these animals developed the ability to glide, they would have been able to reach more distant targets. And as some developed the capacity to generate lift—that is, to actually take off and fly—they further enhanced their range and maneuverability. However, at least as far as we know, none of the theropods that were related to the ancestry of birds was a tree-dweller. How, then, did the capacity for flight arise?

The first possibility is that flight did evolve from arboreal creatures, but that fossils of these tree-dwelling ancestors have not yet been discovered. The second possibility is that flight arose from small, swift, ground-dwelling animals. According to this theory, the ancestors of birds gained an advantage in their pursuit of prey by being able to increase their strides, first by gliding, and then by flying. The answer may long elude us—we must await further discoveries before we can solve the puzzle of the origins of flight.

While dinosaurs, as we usually think of them, died out at the end of the Cretaceous period, the class Dinosauria did not become extinct at that time. It contains more than 9,000 species of living birds, and it remains one of the most diverse groups of vertebrates alive today.

THIS RECONSTRUCTION *of* Deinonychus, *with its characteristic long, low skull, is in the Natural History Museum in London, UK.*

DINOSAUR FEATURES

Bones are the key to our understanding of dinosaurs. Through them, we can identify the features that distinguish dinosaurs from other animals.

Dinosaurs, like all other animals, are identified and united by the common possession of key features. In the same way that we recognize mammals because they have fur or birds by their feathers, we can recognize dinosaurs because they have unique features. However, because all our knowledge of dinosaurs comes from the fossils of their skeletons, their unique features must be found in their bones. And similarly, different groups within the dinosaurs are recognized by their bony features.

LOOKING AT BONES

The skulls of vertebrates contain two bones in the palate called vomers. In dinosaurs, these reach from the front of the snout back to the level of the antorbital fenestrae (holes in the skull in front of the eyes). In most other animals, the vomers are not this long.

Moving down to the shoulder blades (the scapulae), we find that dinosaurs had a socket that faced backward where the arm attached. This feature was perhaps related to another on the upper arm bone (the humerus)—a long, low crest on the upper part that provided attachment for muscles. The hand had a fourth finger that contained no more than three finger bones (phalanges).

The hips were anchored to the spine by three or more vertebrae. The socket where the leg articulated had a hole in the center and an enlarged, bony rim around its upper margin. The ball-like part of the thigh bone (the femur) was turned in toward the midline of the animal. The femur had a bump midway along its length (the fourth trochanter), which was also used for muscle attachment. Farther down the leg, the tibia (shin bone), which had a crest (the cnemial crest), was far larger than the calf bone (fibula). The ankle bone (astragalus) had a process that reached up the leg and fitted into a notch on the tibia. The ankle was a simple hinge, and dinosaurs walked on long toes.

These key features of both the front and back limbs allowed the legs to swing frontward and backward under the animal. This provided a very efficient way of moving, which scientists think may have been one of the reasons that dinosaurs were so successful.

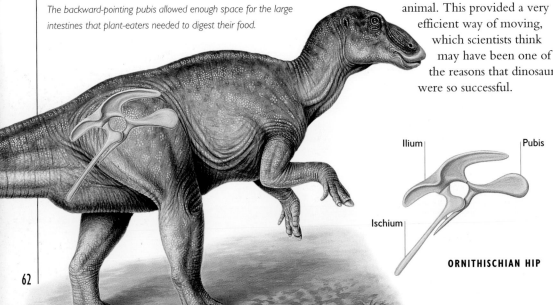

THE HUGE PLANT-EATER *Edmontosaurus had a typical ornithischian pelvis. The backward-pointing pubis allowed enough space for the large intestines that plant-eaters needed to digest their food.*

Ilium | Pubis

Ischium |

ORNITHISCHIAN HIP

BIRDS' HIPS *look similar to those of bird-hipped dinosaurs, but birds evolved from lizard-hipped dinosaurs, which shared common ancestors with modern crocodiles and alligators.*

the others. The neck was long and flexible, curving in an S-shape. These features were modified beyond recognition in some of the more advanced saurischians. In sauropods, for example, the hand developed into an elephant-like foot for bearing the weight of the front of the animal, while in birds the three remaining fingers fused into a complex bone that supported many of the wing feathers.

Ornithischians had only small teeth at the front of the mouth, but these teeth were sometimes lost and replaced by a beak. They also had an extra bone (the predentary) at the front of the lower jaw. This supported the beak. Another ornithischian feature was the development of horns, spikes, plates, frills, and other bone ornaments.

All ornithischians are now extinct, but saurischians survive to this day as birds. As well, all ornithischians were plant-eaters, while saurischians included both herbivores and carnivores.

THE SAURISCHIANS AND ORNITHISCHIANS

Dinosaurs can be divided into two major groups: the saurischian (or lizard-hipped) dinosaurs and the ornithischian (or bird-hipped) dinosaurs. This division was established in 1887 by the English paleontologist Harry Seeley. Some dinosaurs—the meat-eating theropods (including birds), the long-necked sauropods, and the prosauropods—had a pubic bone (one of the major hip bones) that pointed forward in a lizard-like arrangement, while the others had a pubic bone that pointed toward the rear and ran parallel to another hip bone (the ischium). The latter were called bird-hipped dinosaurs because their hips superficially resembled those of birds.

Some other features, too, are unique to each group. Saurischians had a grasping hand with the thumb offset to the other digits and a second finger that was longer than

THE MEAT-EATER Allosaurus *was a saurischian. Its pubis pointed forward between the legs and helped support the leg muscles.*

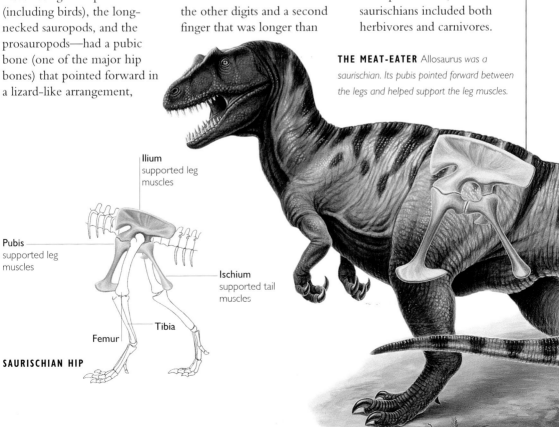

Ilium
supported leg muscles

Pubis
supported leg muscles

Ischium
supported tail muscles

Tibia

Femur

SAURISCHIAN HIP

SKELETONS AND SKULLS

The enormous diversity of dinosaurs as a group can be seen in the great variety of skeletons that have come to light over more than 200 years.

The dinosaurs were a huge and varied group. This is reflected in their skeletons, which show a variety of forms, ranging from the enormous frames that supported the largest animals ever to walk the Earth to the delicate, elegant structures of the smallest dinosaurs. Despite such a diversity of skeletal forms, many common themes were replayed throughout their history.

SKELETAL DIVERSITY
The secret of the dinosaurs' success lies partly in their design. Their legs were held directly under the body, which allowed them to swing forward and backward and

TRICERATOPS *(right) had a huge skull., with a beak for breaking off plant matter and grinding teeth at the back of its mouth.* Tyrannosaurus *could charge in sudden bursts, but its legs (left) were not designed for long chases.*

meant that dinosaurs avoided the sprawling, ungainly gait that is typical of many other reptiles. A number of groups, such as the theropods, also had hollow bones that provided strength and support while keeping the animals' weight to a minimum.

Sauropods also had weight-reducing adaptations in their skeletons. Despite or, more properly, because of their immense size, their skeletons

had to be as light as possible while still providing the strength needed to support their tremendous weight. The neck vertebrae, for example, contained hollows and cavities with many of the processes reduced to struts. Edward Drinker Cope acknowledged this feature when he named *Camarasaurus* ("chambered lizard") for the hollow, boxlike nature of this dinosaur's neck vertebrae.

There were also parts of the skeleton where structural strength was more important than saving weight. Massive, solid leg and arm bones held up the heavy bodies of

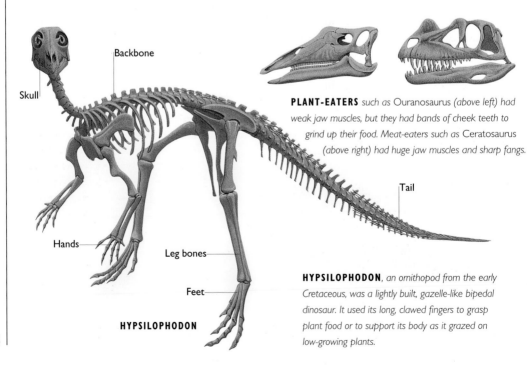

Backbone

Skull

PLANT-EATERS *such as* Ouranosaurus *(above left) had weak jaw muscles, but they had bands of cheek teeth to grind up their food. Meat-eaters such as* Ceratosaurus *(above right) had huge jaw muscles and sharp fangs.*

Hands

Leg bones

Feet

Tail

HYPSILOPHODON

HYPSILOPHODON, *an ornithopod from the early Cretaceous, was a lightly built, gazelle-like bipedal dinosaur. It used its long, clawed fingers to grasp plant food or to support its body as it grazed on low-growing plants.*

sauropods, ceratopsians, stegosaurs, and ankylosaurs. Weight-bearing legs were held as straight as possible, improving their weight-carrying capacity.

Armor, spikes, plates, and shields made of bone tended to make an animal heavy. It is no surprise that these features are found in dinosaurs that moved on all fours, where the extra weight could be distributed more evenly.

In small dinosaurs, the emphasis was on flexibility and agility. Small theropods and ornithopods had lightly built skeletons and a great deal of movement at the skeletal joints. They also tended to have very long legs for their size, indicating that they were fast runners able to duck and dive away from larger predators.

SKULLS

The heads of dinosaurs were encased in bone, which is heavier than flesh or muscle. Because of this, the very large heads of some dinosaurs were extremely heavy.

In the case of ceratopsians, heavy, bony heads were an advantage because the extra bone provided protection

FALSE ALARM

During his dinosaur-hunting days, Jim Jensen, a resident paleontologist at Brigham Young University in Utah, found several important sites—such as the Dry Mesa quarry in Colorado—and discovered and named a number of dinosaurs, including the theropods *Dystylosaurus* and *Torvosaurus*.

During his excavations in Colorado during the 1970s, he uncovered two huge "new" sauropods that he named *Supersaurus* and *Ultrasaurus*. In both cases, however, these dinosaurs were named from only a few bones and some vertebrae. Although these finds created a stir at the time, it now appears that these were just huge individuals of dinosaurs that were already known to scientists—*Diplodocus* and *Brachiosaurus* respectively.

against both attacks from predators and injury during combat with rivals. The mighty weight of a ceratopsian skull was balanced on a short neck and supported by huge muscles.

The heads of sauropods, on the other hand, were perched at the end of long necks.

THE HADROSAUR *Corythosaurus (above left) had a crest on its skull and nostrils at the front of its snout. Like all sauropods, Brachiosaurus (above right) had its nasal openings at the top of the head, above the eyes.*

These heads had to be big enough to permit the animals to collect sufficient food but small enough not to weigh down the neck. In some sauropod skulls the weight was minimized by expanding holes—or "fenestrae"—in the skull, thus reducing the bone to thin struts and rods.

Theropods generally had large heads, which not only had to be as light as possible but also had to withstand the tremendous forces that were transmitted through the head when the animal bit into its prey or while it held onto another animal in a violent struggle. Theropods' skulls, therefore, also had large holes, but the bony struts that surrounded them were still very solid.

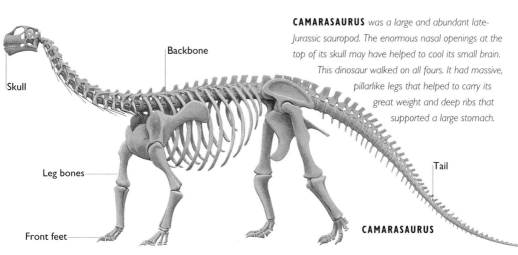

CAMARASAURUS *was a large and abundant late-Jurassic sauropod. The enormous nasal openings at the top of its skull may have helped to cool its small brain. This dinosaur walked on all fours. It had massive, pillarlike legs that helped to carry its great weight and deep ribs that supported a large stomach.*

Backbone

Skull

Leg bones

Front feet

Tail

CAMARASAURUS

WARM- OR COLD-BLOODED?

Until quite recently, it was generally accepted that the dinosaurs, like present-day reptiles, were cold-blooded animals. Recent research has raised serious doubts about this assumption.

One of the most lively scientific debates about dinosaurs in recent times has been about whether these animals were, as had long been assumed, cold-blooded creatures, or whether, like mammals and birds, they were warm-blooded. This debate has raised important questions about dinosaur physiology. We may never get a definite answer to this puzzle, for the very reason that, unlike bones, body temperature does not fossilize. However, we can make some inferences about the body temperature of dinosaurs and their capacity to regulate it by looking at the fossil evidence and using our knowledge of living creatures.

Actually, the question is not about whether a dinosaur's blood was warm or cold. The terms "warm-blooded" and "cold-blooded" are commonly used, but they are misleading. On a hot day, for example, a cold-blooded animal such as a crocodile

IN 2000, COMPUTER- *enhanced images of the late-Cretaceous dinosaur* Thescelosaurus *revealed what looked like a fossilized heart. Some scientists think it has similarities to that of a mammal, lending weight to the theory that dinosaurs were warm blooded. Others think the "heart" is simply a concretion—a deposition of minerals.*

may have a higher blood temperature than a warm-blooded mammal of a similar size. What the debate is really about is this: Were dinosaurs able to maintain a constant body temperature, as birds and mammals do, or did their body temperature, like that of lizards, snakes, and crocodiles, fluctuate in response to the environment?

There are advantages

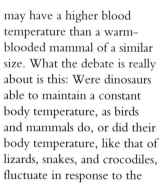

THE LARGE SAIL on *the back of* Ouranosaurus *may have enabled this big plant-eater to regulate its body temperature.*

and disadvantages in being able to maintain a constant body temperature. The main advantage is that an animal is always ready for action—night or day, hot or cold. This means that it is able to exploit habitats and endure conditions, such as near-freezing nights, that could not be tolerated by animals whose body temperature is very much dependent upon the environment. The main drawback of maintaining a constant body temperature is that it requires a great deal of energy. As a result of this, an animal needs to eat more. A warm-blooded lion, for example, must eat about ten times as much as a similar-sized cold-blooded crocodile.

There are convincing arguments to suggest that dinosaurs displayed a variety of physiologies—that some could be recognized as truly warm-blooded while others were cold-blooded.

WARM-BLOODED ANIMALS *such as mice can be active all the time. Crocodiles and other cold-blooded animals must take in heat in order to be active.*

THEROPODS

Theropods were the ancestors of the warm-blooded birds and this has suggested to some researchers that they would have been warm-blooded. In addition, recent dinosaur finds in China show that small theropods had "hairy" coats, or even feathers, that would have helped trap heat inside the body. Smaller theropods that weighed less than about 220 pounds (100 kg) could certainly have benefited from maintaining a constant body temperature, as this would have helped to keep them active while hunting. Several of these theropods, such as the dromaeosaurs and the troödontids, had slashing claws that are more typically associated with active warm-blooded creatures rather than with low-energy, cold-blooded animals.

Another argument for warm-blooded dinosaurs cites the small theropods and small ornithopods that have been found at high-latitude sites such as Dinosaur Cove in Victoria, Australia, where Mesozoic winter temperatures were below freezing. Today, such environments are the domain of warm-blooded animals. However, making analogies between the past and the present can be misleading. Remains of other typically cold-blooded Cretaceous animals such as crocodiles and large amphibians have been also found at these once-chilly sites.

EVIDENCE FROM CROCODILES

Recent studies of body temperature in crocodiles are relevant to the question of temperature regulation in larger dinosaurs. Crocodiles lose heat to the environment more slowly as they get larger. In other words, the increasing bulk of the crocodile acts to trap heat within the body. A hypothetical 5-ton (5.1-t) specimen could maintain a constant body temperature just by being large. If this heat-trap system applied to dinosaurs, the largest of them would have needed to get rid of heat absorbed from the environment or generated by moving muscles and digesting food. Long necks and tails, such as those of sauropods, would have helped because they increased the surface area relative to the mass of the animal. Plates, spikes, and sails could also have helped drain heat from the body. This may explain the function of the plates of the stegosaurs or the sails on the backs of *Spinosaurus* and *Ouranosaurus*. Such features would have been useful to large dinosaurs living in warm environments.

DROMICEIOMIMUS *was a small, extremely active predator. It is thought that it may have been warm-blooded.*

SURVIVING IN THE COLD

Husband-and-wife team Tom Rich and Patricia Vickers-Rich migrated from the United States to settle in Victoria, Australia. They are key players in Australian vertebrate paleontology and have made many finds in Mesozoic and Tertiary deposits.

Their most famous site is Dinosaur Cove on the southern coast of the Australian mainland. Here they dug a mine into a sea cliff, following the bed of a Cretaceous stream. Their efforts were rewarded by several dinosaur finds. These included *Leaellynasaura*, a small ornithopod named for their daughter, and *Timimus*, a tiny ornithomimosaur named for their son. In another Cretaceous site, they have found some of Australia's oldest mammal fossils.

Dinosaur Cove has turned out to be the most productive dinosaur site in Australia, and it provides hard evidence that dinosaurs could live in very cold environments. *Leaellynasaura* had huge eyes that would have helped it to see in the extended nights of its high-latitude home. It was too small to migrate to a new area each year so, somehow, it must have survived winters in which temperatures plummeted well below freezing.

SIGNIFICANT FOSSILS WERE DISCOVERED AT THIS LOCALITY, DINOSAUR COVE, IN 1980. FIELD PARTIES COMPOSED PRINCIPALLY OF VOLUNTEERS FROM MONASH UNIVERSITY, THE MUSEUM OF VICTORIA AND EARTHWATCH COLLECTED DINOSAURS AND OTHER VERTEBRATE FOSSILS FROM THREE SITES WITHIN THIS COVE, 1984 - 1993. MAJOR SUPPORT CAME FROM THE NATIONAL GEOGRAPHIC SOCIETY, ATLAS COPCO, I.C.I., THE DEPARTMENT OF CONSERVATION & NATURAL RESOURCES, AND THE AUSTRALIAN RESEARCH GRANTS COMMITTEE.

THE BRAIN AND SENSORY SYSTEMS

Recent research has helped dispel many persistent myths about dinosaur brains and intelligence and about how these animals experienced their world.

Like most soft tissue, brains, nerves, and the sensory systems of dinosaurs do not fossilize. But we can learn a lot about what they were like by studying the bones that encased the brains, eyes, and ears as well as the various passages that held nerves.

The brain is surrounded by bones that approximate the size and shape of the living brain. The hole that is left by the brain can be filled with matrix or plaster. This produces an endocast—a copy of the brain space in the skull. One problem with this technique is that it requires pulling the bones apart to get at the endocast, a destructive procedure that could result in irreparable damage to the precious skull. Modern

THIS CAT SCAN *of a Tyrannosaurus braincase is in the Field Museum in Chicago, USA.*

technology has helped here in the form of Computer Assisted Tomography (CAT) scans where the skull can be X-rayed in three dimensions and the brain space analyzed on computer.

MISCONCEPTIONS

Ever since Othniel Charles Marsh described the first known endocasts in 1896, it was believed that dinosaurs had relatively small brains for their size. This is, in fact, a misconception. In living animals, brain size does increase

with body weight, but to a much smaller degree. The difference in size between a particularly large animal's and a much smaller animal's brain is only slight in relation to the difference in body size.

Taking this into account, dinosaur brains, although small, were proportionally the same size as those of modern reptiles or amphibians. Some small theropod dinosaurs had relative brain sizes comparable to that of living ratite birds such as ostriches and emus. However *Stegosaurus* had the smallest brain-to-body size ratio of any known terrestrial vertebrate.

Another misconception that characterized popular ideas about dinosaur brains is that some dinosaurs had "helper" brain in their hips. This notion dates back to Marsh's work on *Stegosaurus*. Marsh noticed an enlarged space in the hips that could have accommodated a swelling of the spinal cord. Such a ganglion could have helped relay messages to the rear of the animal or even controlled the posterior parts, freeing up the tiny brain in the head for other cognitive functions. However, this rather fanciful idea no longer holds sway. Scientists now

A MALE PARASAUROLOPHUS

had a huge head crest in which air moved along a complex series of passages. There are several theories about the function that this crest performed. It seems likely, however, that it may have served as an olfactory organ.

STUDYING DINOSAUR INTELLIGENCE

Dale Russell is a Canadian paleontologist now working in the United States. He specializes in theropod dinosaurs and has named several dinosaurs, including *Archaeornithomimus*, *Daspletosaurus* and *Dromiceiomimus*.

Much of his earlier work involved studies of dinosaur brains and endocasts. One of his most controversial ideas is of a hypothetical intelligent dinosaur that he christened the "dinosauroid." Russell reasoned that, because toward the end of the Cretaceous small theropod dinosaurs were developing relatively large brains (for dinosaurs), they may, if they had been able to follow this trend for a few more tens of millions of years, have evolved into an intelligent species that went on to dominate the planet. Only ever intended as a thought experiment, it is indeed food for thought!

believe that the swelling was probably filled with tissues other than nerves.

SENSES

Like the brain, the eyes, ears, and nasal cavities of dinosaurs were surrounded by bones, and a study of these helps us to understand the nature of sensory organs.

Theropods generally had well-developed eyes that included a ring of bone within the eye (the sclerotic ring). As well, the part of the brain that dealt with vision was enlarged. This suggests that these dinosaurs relied on sight as a primary sense for locating prey. Some plant-eaters, such as the ornithopods and pachycephalosaurs seem

to have relied on a keen sense of smell to detect predators at a distance. The small ornithopod *Leaellynasaura* had enormous eyes and optic lobes in the brain that probably helped it see in the extended Antarctic nights. Small theropods probably had a refined sense of balance because the area of the brain that deals with balance is enlarged in these animals.

INTERPRETING FOSSILS

Interpreting fossil structures in bones devoid of flesh requires skilled deductive reasoning. This is particularly the case in structures such as the nasal cavities of ankylosaurs and hadrosaurs. Both developed convoluted pathways for the air passages as they passed

through the skull. The most extreme example of this is *Parasaurolophus*, where the complex air passages extended the full length of a head crest that was over 3 feet (1 m) long. These structures could have been sounding instruments—long tubes that made deep, resonant tones that helped the animals communicate. Another theory holds that the passages were lined with olfactory cells that would have given their owners a very keen sense of smell. Yet another theory maintains that they served to bring warm air into the lungs, or to trap moisture from exhaling air, or even that they helped to keep the brain cool. We may never know which, if any, of these theories are correct, but it could well have been that extended air passages had several functions.

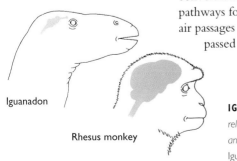

Iguanadon

Rhesus monkey

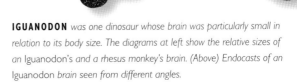

IGUANODON *was one dinosaur whose brain was particularly small in relation to its body size. The diagrams at left show the relative sizes of an* Iguanodon's *and a rhesus monkey's brain. (Above) Endocasts of an* Iguanodon *brain seen from different angles.*

FEEDING AND DIGESTION

Dinosaurs employed a wide range of strategies,

both to gather or catch their food and then

to digest and process it.

Dinosaur diets can be inferred from a number of lines of evidence. Teeth, claws, and jaws indicate food preference while the occasional fossil with preserved stomach contents provides direct evidence of what a dinosaur was eating. Coprolite (dung) fossils and an understanding of the surrounding flora and fauna enable us to make reasonable assertions about which dinosaurs were eating which types of foodstuffs.

CARNIVORES

Theropods typically had long arms with sharp, curved claws that allowed them to grab their prey and rip at the flesh with rows of slashing teeth. The large Jurassic theropod *Allosaurus* probably hunted animals ten times its size. It may have attacked these huge beasts by ambushing them, slashing at them, then withdrawing until the prey

ALLOSAURUS, *a theropod, had huge jaws and teeth like steak knives.*

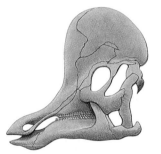

CORYTHOSAURUS *had a horny beak for stripping leaves off plants.*

was weakened by blood loss. *Allosaurus* may also have fed on the young of large sauropods or attacked more modest-sized stegosaurs and camptosaurs. The smaller theropods had flexible skeletons that would have allowed for greater agility.

PROTOCERATOPS *sheared off plant matter with its sharp beak.*

IGUANODON *used grinding teeth to crush plant matter into a pulp.*

ANATOTITAN *was a plant-eater with about 1,000 tiny leaf-shaped teeth.*

They either chased after smaller animals or formed groups to attack larger prey.
    The exception among the theropods were the tyrannosaurs, which had puny arms.

SAUROPOD STONES

David Gillette is an American paleontologist whose work has focused mainly on the Jurassic Morrison Formation in New Mexico. One of the specimens he excavated there was the giant skeleton, more than 140 feet (43 m) long, of the late Jurassic sauropod *Seismosaurus*.
    During the excavation of *Seismosaurus*, which took eight years, Gillette noted and recorded the exact position of more than 240 gastroliths, most of them the size of apples, that were associated with the specimen. Most of these were collected in an area of the rib cage just in front of where the stomach would have been, indicating the presence of a huge gizzard. Others scattered through the specimen signposted other parts of the animal's digestive tract.

These seem to have been of no use in capturing prey. However, tyrannosaurs had particularly large mouths and the most powerful bites of any known animal, past or present. As well, they were the largest animals in their habitat, so that an individual tyrannosaur could tackle any potential prey.

HERBIVORES

Plants that grew throughout most of the Mesozoic were both poor in nutrients and relatively hard to break down. Plant-eaters employed a number of strategies to deal with these problems, most of which involved processing large quantities of food.

Sauropods stripped and swallowed plant matter largely without processing it in the mouth. While they could thus take in vast amounts of food, they had to break it down to retrieve the scarce nutrients. They appear to have done this in a huge, vatlike pre-stomach, or gizzard, where the incoming food could be stewed and brewed into a nutrient soup. This process was helped by "gizzard stones," or gastroliths, that were held in the gizzard and helped stir up the brew.

The pachycephalosaurs, ornithopods, and ceratopsians employed a different strategy. They used grinding teeth to break down the food before swallowing it. This process took its toll on the teeth, which quickly wore down. However, a number of times throughout their lives these dinosaurs were able to discard their worn-out teeth and grow new ones. The advanced ornithischians took this technique to its limit, evolving batteries of tightly packed teeth that functioned like a single grinding plate. These

DUNG FOSSILS, or *coprolites, probably from the plant-eater* Titanosaurus, *display remnants of plant matter.*

batteries grew continuously through the animal's life and could contain hundreds of teeth. The teeth at the front of the mouth snipped off the plant material for grinding in the back of the mouth.

Ornithischians lost their front teeth relatively early in their evolution, replacing them with a sharp, birdlike beak. These animals appear to have had cheeks that prevented food from falling out while they chewed. Ridges of bone around the mouth that probably were supports for cheeks have been observed in several of these dinosaurs.

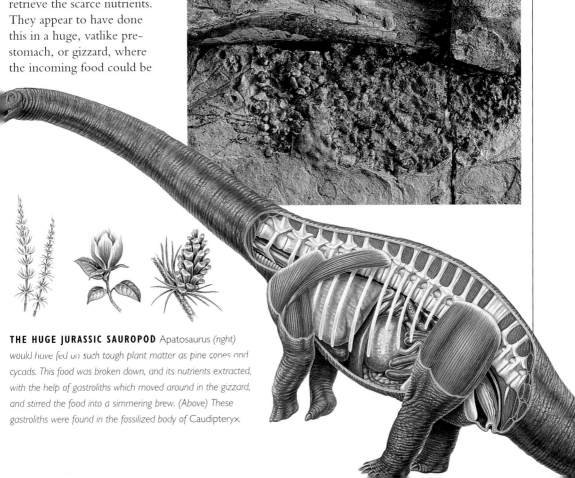

THE HUGE JURASSIC SAUROPOD Apatosaurus *(right) would have fed on such tough plant matter as pine cones and cycads. This food was broken down, and its nutrients extracted, with the help of gastroliths which moved around in the gizzard, and stirred the food into a simmering brew. (Above) These gastroliths were found in the fossilized body of* Caudipteryx.

REPRODUCTION AND THE LIFE CYCLE

The study of dinosaur egg fossils, which dates back to the 1920s, has added greatly to our understanding of dinosaur reproduction and growth patterns.

As with all animals throughout history, dinosaur reproduction relied on mating between the sexes. While it is often very difficult to determine which gender a particular dinosaur skeleton belonged to, there are a number of cases where skeletons of a single species fall into two distinct forms. It is reasonable to assume that these differences were, in some way, associated with courtship, nesting, and the rearing of offspring.

The frills of ceratopsians, for example, are more likely to have played a part in courtship behavior than in defense. The frill of the small ceratopsian *Protoceratops* is, in places, eggshell thin and would have offered little protection against predators.

A MAIASAURA *embryo at an advanced stage of development. The yolk sac, shown here in yellow, provided nourishment for the unhatched juvenile.*

However, roughly half of the hundred or so known *Protoceratops* skulls have a broad neck frill while the others have a less extensive frill. It is possible that the male had the larger frill, and that it may have been brightly colored or used in some other way to impress females or to intimidate rival males. There are several skulls of ceratopsians that show signs of damage from the horns of another ceratopsian. This has been construed by some scientists as direct evidence that fierce rutting took place between bull males that were vying for the right to copulate with females.

There are two kinds, or morphs, of *Tyrannosaurus*—a heavy form and a lighter, more gracile form. Evidence from the tail bones indicates that the smaller, gracile form is the male. This seems counter to our intuition because, in humans, males tend to be larger. But in many animal species females are larger and this usually confers advantages in egg production or defense of the young.

Among the hadrosaurs there are differences in head crests that also suggest some differences between the sexes.

THIS CAMPTOSAURUS *has reached the end of its life cycle, killed either by disease or old age. Small compared to most plant-eaters, it still probably had a life span of at least several decades. The small theropods, Coelurus, which here lurk in the background, would have lived for a much shorter period.*

ROY CHAPMAN ANDREWS *(right), from the American Museum of Natural History, and his assistant, George Olsen, are seen here excavating a nest of about 20 Protoceratops eggs at Bayn Dzak, at the foot of the Flaming Cliffs in the Gobi Desert, Mongolia, in the early 1920s. Both skeletons and eggs of Protoceratops andrewsi, which was named after the famous American paleontologist, were discovered during these early expeditions to Mongolia. Until then, this small, primitive, horned dinosaur had been unknown to science.*

In *Parasaurolophus*, for example, half the known adult specimens had a long crest, while in the rest the crest was much shorter. Perhaps this feature served as a courtship device that allowed the males to boom out a mating call deeper and more striking than the calls the smaller-crested females were capable of.

HOW MATING WAS ACHIEVED

Once mates had been found it was time to copulate. Internal fertilization is a common strategy for terrestrial animals as it prevents damage to the sperm. We can, therefore, reasonably assume that some kind of internal fertilization occurred with dinosaurs and that copulation was necessary. Male dinosaurs almost certainly had penises, because their living relatives, the crocodiles, have penises. Some birds, too, have penises, although most now do not.

We do not know exactly how dinosaurs copulated. For all dinosaurs, the possession of a prominent tail would have been a significant obstacle to copulation. Some modern male animals with similar tails, such as lizards and crocodiles, have two, sideway-facing "hemipenes" that allow lateral entry into the female. Snakes copulate by wrapping around each other in a spiral, in this way bringing their genital openings together.

Besides tails, there were two other significant barriers to copulation in some dinosaurs. The massive size of the larger dinosaurs suggests that it would have been very difficult for a male to mount a female and that doing so would have placed huge strains on the rump and hind legs of the female. A male-on-top position is particularly problematic in stegosaurs, where plates and spikes along the back would have presented extra and unavoidable complications for copulation.

Dinosaurs had a significantly different reproductive strategy from modern mammals. In modern mammals with a size range similar to dinosaurs, a few, well-developed offspring are born at a time and they tend to survive reasonably well to adulthood. Dinosaurs, however, appear to have laid lots of eggs during a season and, despite varying degrees of parental care, the survival rate of the young into adulthood seems to have been rather low. Effectively, dinosaurs relied on the quantity of offspring for the perpetuation of a species. In contrast, modern large mammals rely on carefully nurturing a small number young, in order to preserve their kind.

Adult male head   Adult female head

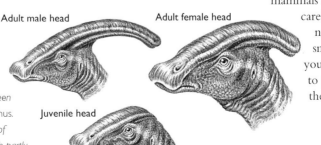

THREE CREST SIZES *have been found in fossils of Parasaurolophus. They probably represent adults of different sexes and juveniles with partly formed crests.*

Juvenile head

A CLUTCH *of fossilized Troödon eggs (left). A fossilized embryo of Oviraptor (below left) and skull fragments of a baby dromaeosaur (below right) are part of the collections of New York's American Museum of Natural History.*

NESTS AND BABIES

Most of what we know about dinosaur reproductive behavior has come to light only in the last few decades, when the study of dinosaur nests, nest sites, eggs, and young became reinvigorated. The first dinosaur eggs were discovered in France in 1859, and the first recognized nests were uncovered in Mongolia in the 1920s. It was not until nest sites were uncovered in Montana in the late 1970s, however, that the study of dinosaur reproductive behavior received a significant boost. We now know of more than 200 sites located all over the world that have dinosaur eggs, nests, babies, or the footprints of young.

Despite popular myths and misconceptions, dinosaur eggs were not really huge—the largest known egg is about the size of a cantaloupe. They varied in shape from spherical to an elongated ellipsoid and had differing surface textures, including pitting, lumps, bumps, and raised lines.

Dinosaur remains reveal a wide variety of behaviors relating to nesting and the rearing of offspring. Nests have been found in shallow pits that were filled with vegetation, which would have kept the eggs warm as it decomposed. In other cases, eggs seem to have been scattered around with no attempt at building a nest. In some sites, the arrangement of eggs within the nest indicates that the parents had moved the eggs into specific patterns.

Dinosaurs seem to have had favored nesting areas that they revisited year after year. Nests have been found to be isolated, spread out in a haphazard arrangement, or, concentrated in sites where dinosaur rookeries formed, tightly packed together, and separated by distances that are equivalent to the length of the adult animal. In places such as Egg Mountain in Montana, where the famous *Maiasaura* nests were discovered in 1978, it appears that hundreds, possibly even thousands, of dinosaurs congregated into huge dinosaur rookeries during the breeding season.

Some dinosaur parents may have stayed with the nest while the eggs developed, as most birds do today. Such behavior was illustrated very graphically in 1996 when a fossilized adult *Oviraptor* was found still huddled over its fossilized eggs in a posture identical to that of modern brooding emus and ostriches. It also seems likely that other dinosaurs laid eggs and then, as many modern lizards and turtles do, abandoned them. These young would have had to fend for themselves after they hatched. This may have

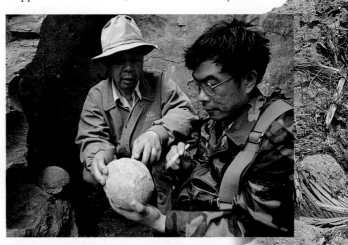

CHINESE PALEONTOLOGISTS *inspect a fossilized dinosaur egg discovered in Hubei province.*

been the pattern for the large sauropod dinosaurs, where the small hatchlings could easily have been trampled underfoot by their gargantuan parents.

LEAVING THE NEST

Fossils indicate that different types of dinosaurs hatched from their eggs at different stages of development. Sauropod hatchlings were relatively advanced in their development and capable of independence. Hadrosaurs, though, were not particularly well developed and resilient when they hatched. Crucial joints in their limbs had not fully formed, and so the tiny hatchlings were dependent on their parents throughout the first weeks or months of life. In some nests, juveniles that were more developed than their siblings, and whose teeth showed signs of wear, have also been found. This adds weight to the theory that some dinosaur chicks stayed in the nest and were fed by attendant adults.

Dinosaurs do appear to have matured at quite a considerable rate.

For example, a hatchling hadrosaur—perhaps 10 inches (25 cm) long—could grow to over 6 feet (2 m) in a few years and attain an adult length of 26 feet (8 m) in a decade. Dinosaurs appear to have grown rapidly in their earlier years, but their growth rates slowed dramatically at adulthood. Some dinosaurs effectively stopped growing at an adult size, while others may have continued to grow steadily throughout life.

Life spans for dinosaurs are hard to calculate. They may have been as short as four or five years for small ones such as *Troödon*, but as long as 150 years for large sauropods.

Be a good animal,

true to your animal

instincts.

The White Peacock,
D.H. LAWRENCE (1885–1930),
English novelist

EGG MOUNTAIN

John "Jack" Horner is a largely self-taught paleontologist, who works mainly in Montana. His most famous discovery followed a visit with a friend to a fossil shop in the small Montana town of Choteau, where they saw tiny, fossilized bones that they recognized as belonging to baby dinosaurs.

Following directions given by the shop owners, Horner soon found what has become known as "Egg Mountain," the nesting ground of the hadrosaur *Maiasaura*. Although dinosaur nest sites and eggs had been found before, this was the first time that a systematic study of dinosaur nests was possible.

Among the numerous startling and fascinating findings of John Horner's research was the very birdlike nature of a dinosaur rookery and the previously unsuspected degree of dinosaurian parental care.

THIS RECONSTRUCTION *of a Maiasaura nest is modeled on the remarkable "Egg Mountain" site in Montana.*

DINOSAURS IN MOTION

Dinosaurs moved in many different ways. Some were four-footed giants that ambled along; others were small, agile bipeds capable of spectacular bursts of speed.

One of the keys to the success of the dinosaurs was their posture. Even in the largest quadrupedal dinosaurs, the bulk of their weight was carried by the hind legs. For many dinosaurs, this meant that they could rise up onto their hind legs without too much effort, a stance that would have been very useful, either for defense or to display to members of their own species. In other dinosaurs it meant two modes of carriage: all fours for energy-efficient movement across the landscape; or two legs for fast getaways. Many dinosaurs were entirely bipedal. This meant that their arms and hands were free for grabbing prey or handling food.

The ancestral dinosaurs were all bipedal, and several groups that were quadrupedal evolved this feature independently. The ancestral dinosaurs bequeathed another asset to their descendants: They carried their hind legs directly under the body—held straight vertically and moving backward and forward in the same plane as the body and the direction of travel. Most animals that competed with the dinosaurs had a more sprawling gait— their legs were held out to the sides and they moved by sweeping their legs around in the direction of travel. This

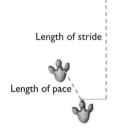

Length of stride

Length of pace

 Length of foot

was more cumbersome—and it required more energy— than the streamlined movements of the dinosaurs.

FAST AND SLOW MOVERS

Standing on two long back legs, some dinosaurs, such as the ornithomimids, could sprint at considerable speed. The speed at which a dinosaur could move along can be measured partly by comparing the length of the leg below the knee (the shin) with the length of the leg above the knee (the thigh). The longer the shin, the wider the stride; and the longer the thigh, the more powerful the forward thrust. Ornithomimids and oviraptors had particularly long shins and relatively short thighs—even though their thighs were quite long in

DINOSAUR TRACKWAYS *provide three basic clues to the the speed at which the dinosaurs were moving.*

relation to the animals' overall size. Estimates of just how fast these dinosaurs could move are derived from theoretical calculations that are based on the skeleton and from careful measurements of fossilized trackways. It would appear that fleet-footed theropods and ornithopods could reach speeds of at least 25 miles per hour (40 km/h).

In general, the larger an animal, the more slowly it moves. Large theropods could probably move no faster than 9 miles per hour (15 km/h).

Four-legged dinosaurs were still constrained by the "larger means slower" rule, but the ability to gallop and trot offered gaits that could be sustained for longer periods.

For very large dinosaurs such as the sauropods, speed probably was never an option. Their sheer weight limited them to slow ambling. Their legs were massively built, and the thoracic and lumbar vertebrae were arranged in an arc that supported the weight of the body in a construction that was similar to that of a suspension bridge.

STRUTHIOMIMUS, *with its long, birdlike hindlimbs, similar to the legs of an ostrich (above), was a very swift runner.*

THE AMERICAN PALEONTOLOGIST
Barnum Brown is seen here marking out a group of fossilized theropod footprints.

DINOSAUR HERDS

Trackways record how dinosaurs moved. They also supply us with unique information about dinosaur groupings and the structure of herds.

Larger theropods seem to have been solitary creatures, or they moved about only in couples or in small groups. The smaller theropods favored larger groups numbering tens of individuals. In at least one example of fossilized tracks that is located near Winton, in Queensland, Australia, dozens of small theropods gathered with a similar number of small ornithopods in a mixed group, only to be set to flight by the appearance of a lone, large, predatory theropod.

Sauropods appear to have moved in groups of about a dozen individuals of the same species—but of different ages. There is evidence that the younger animals kept to the center of a group, while older animals formed a protective cordon around them.

Larger ornithopods and ceratopsians formed vast herds numbering hundreds, or perhaps thousands. As such large groups of big herbivores would quickly have stripped areas of their plants, they must have survived by migrating annually to seasonal pastures. This theory is supported by dense trackways that show hundreds of animals moving in the same direction and fossils of the same species that have been found in distant locations. In many respects, a migrating herd of ceratopsians or hadrosaurs must have resembled modern migrations of buffalo in North America or wildebeest in Africa.

A PAIR OF SMALL PLANT-EATERS *(right foreground), alarmed by the appearance of two predatory theropods (below), break away from protection offered by a migrating herd of long-necked sauropods as it moves across the North American Jurassic landscape.*

METHODS OF SURVIVAL

All dinosaurs, both carnivorous and herbivorous, faced challenges to their survival. They evolved a number of strategies to help them find food and escape attack from predators.

Life for dinosaurs, as for all animals, consisted largely of finding enough food to survive while avoiding becoming a meal for another creature. Herbivores evolved complex behavioral strategies and an array of defensive armaments to protect themselves from attacks by predators. Locked into an evolutionary arms race, and often facing keen competition for available prey, predators evolved ever-more deadly weapons that they could bring to bear in attacking and subduing the animals they preyed upon.

SAUROPOD STRATEGIES
For the largest dinosaurs, their sheer size provided an effective defense against most predators. Huge sauropods were probably impregnable against the attacks of smaller predators. But the young sauropods were not protected by size and would have been easy prey. Defense for these animals probably came in the form of social structures, where adults could fend off attacks against their young

by shielding them with their vast bodies.

The tail of an adult sauropod was a weapon that could be wielded like a baseball bat and slammed into an attacker. The slender, more whiplike tails of diplodocids could deliver deadly, stinging blows.

BUILT-IN PROTECTION
The horns of ceratopsians and the tail spikes of stegosaurs could have been brought to bear with devastating effect against predators. These bony structures had a covering sheath of keratin and formed a lethal, piercing point. The spikes that covered ankylosaurs may also have been used as offensive weapons. *Edmontonia*, with its skull protected by plates,

THE HEAD *of Triceratops (left) was equipped for both attack and defense. (Above) The sharp, serrated teeth of Daspletosaurus.*

could have driven its large, sharp, forward-pointing shoulder spikes into an attacker with crippling effect.

An ankylosaur had a club on the end of its tail. This stiffened into a bony mace. Swung from side to side, it became a weapon that would have been most effective in cracking the delicate ankles of theropod predators. Ankylosaurs were also protected by a sheath of bony armor that could have withstood bites and slashes from sharp-toothed and sharp-clawed predators.

Display could also be a useful defensive weapon. The frills of ceratopsians and the plates of stegosaurs may have been capable of flushing with color when the animal was aroused or felt threatened. This display, which also had the effect of making the animal look bigger and more intimidating

EUOPLOCEPHALUS *had studded, bony body armor and a club on its tail.*

FINDS IN THE GOBI DESERT

The Polish paleontologist Zofia Kielan-Jaworowska studied paleontology at the University of Warsaw and later became director of the Institute of Palaeozoology in Warsaw. She headed a series of three joint Polish and Mongolian expeditions to Mongolia between 1963 and 1971.

She and her colleagues found many new sites across the Gobi Desert and made many new finds of various types of dinosaurs, including ornithomimosaurs, pachycephalosaurs, sauropods, and theropods. They also discovered the fossil remains of numerous mammals and other animals.

In 1965, in late Cretaceous rocks of the Nemegt Basin, her team unearthed the terrifying claws and arms of *Deinocheirus* and in 1971 they made one of the most famous of all dinosaur fossil finds: a *Velociraptor* and a *Protoceratops* locked in mortal combat, both apparently killed when a sand dune collapsed on them.

than it really was, could have persuaded many a would-be predator to abandon its attack.

For many larger ornithopods, safety in numbers was probably their main defense. Herding in tens, hundreds, or even thousands of individuals would have given predators a difficult target to attack. Large herds meant that there were more eyes and ears to seek out out possible attackers, and greater opportunities for alerting others in a herd to an impending attack.

ON THE ATTACK

Predatory dinosaurs needed a suitable armory of weapons with which to press home their assaults. These weapons were probably both behavioral and structural.

Some theropods were probably ambush predators. Waiting unseen beside trails, they were ready to pounce on prey when the opportunity presented itself. Others could have hunted alone and in the open, their size and strength being the decisive factors in any combat.

Pack hunting was an effective tactic for predators smaller than their intended prey. Some dromaeosaurs may have hunted in packs. The fossils of several *Deinonychus* found with a single *Tenontosaurus* could be construed as evidence of cooperative behavior in these hunters. It is also possible that small groups of allosaurs cooperated in attacks on sauropods—some distracting the attention of adults while their accomplices attacked and killed younger individuals.

Theropods were well-armed with cutting and slashing teeth and long, curved claws that could grip or tear at their victims. Some claws, such as the huge

THIS DRAMATIC RECONSTRUCTION *shows the small and very agile theropod* Dromaeosaurus *launching an attack with its teeth and sickle-like claws on the largest of the hadrosaurs,* Lambeosaurus.

hand claws of *Baryonyx* or the thresher claws of *Therizinosaurus*, appear to have grown to extraordinary lengths and probably had very specific functions. Similarly, the slashing toe claws of the dromaeosaurs and the troödontids were highly specialized and perfectly adapted for maiming or killing prey.

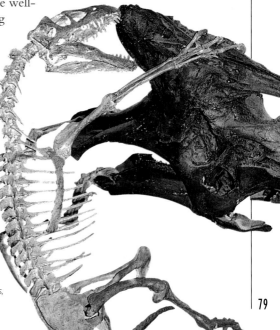

THE PLANT-EATERS

For most of the age of dinosaurs, the climate was warm and moist. Plants grew in abundance, feeding hundreds, or perhaps thousands of species of dinosaurs.

Plant-eating dinosaurs came in an extremely wide range of sizes and designs. There were both the ornithischian and saurischian plant-eaters, as well as quadrupedal and bipedal ones. In fact, as far as we know, all the ornithischian dinosaurs were plant-eaters.

THE SAURISCHIA

There are two categories of saurischian plant-eaters—the prosauropods and the sauropods. The earlier pro-sauropods were medium-to large-sized dinosaurs with long necks and tails, relatively small heads, and large bodies. They were all quadrupedal, although some may have been capable of rocking up onto their hind legs for feeding, defense, or display.

Sauropods probably had a prosauropod ancestor and ranged in size from large to extremely large. Some were the largest animals ever known to have walked the Earth. They are instantly recognizable by their very long necks and tails, large barrel-like bodies, and very small heads.

Both the sauropods and the prosauropods relied for their survival on processing large quantities of low-quality food. They achieved this by rapidly stripping leaves and fronds that were swallowed whole, without chewing, and

were broken down by the process of fermentation in their huge, vatlike stomachs. Gizzard stones, or gastroliths, have been found in the stomachs of many sauropods. These stones were ground to-gether by the muscular action of the stomach and helped to crush the very tough fibers of plant matter.

THE ORNITHISCHIA

While ornithischian dinosaurs represent a great diversity of dinosaur types, all were plant-eaters. Within this group are the shielded thyreophorans, including stegosaurs and ankylosaurs, the ornithopods, including the iguanodontids and the hadrosaurs, and also the marginocephalians, which include the ceratopsians and the pachycephalosaurs.

All thyreophorans had some form of bony armor on their backs. In stegosaurs, this consisted of two rows of bony plates and spikes that were held vertically from the body.

LONG NECKS *enabled sauropods (left) to reach the foliage at the tops of trees. The teeth of Othnielia (far left) were completely covered with protective enamel.*

In ankylosaurs, the back was covered in multiple rows of bony lumps that sometimes extended onto the flanks or even onto the belly. Some of these bony lumps, particularly those that were positioned around the edges, could develop into spikes. The armor of stegosaurs and ankylosaurs weighed them down heavily onto their four legs, but some of the earlier thyreophorans, such as *Scutellosaurus*, may have been

able to get around on two legs for short periods.

Ornithopods take their name (meaning "bird-foot") from the three-toed, birdlike feet of many members of the group. They varied in size from the diminutive *Hetero-dontosaurus* to hadrosaurs that were 30 feet (10 m) or more in length. The group was characterized by animals that had relatively large heads, moderately long necks, and long hind legs. They could travel on all four legs or rise up on two if more speedy or agile movements were required. There is also a trend in this group of dinosaurs to develop bony struts ("ossified tendons") along the back, over the rump, and down the tail. These probably helped to hold the tail and the rear of the animal steady, reducing any flexing, and thus helping to control some movements.

The marginocephalians were a group of dinosaurs that featured some form of bony growth around the margin of the head. Pachycephalosaurs sported a series of lumps and

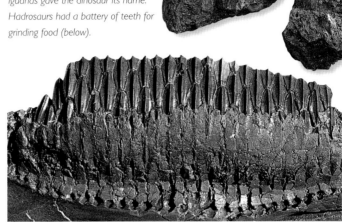

THE FIRST FOSSILS *of* Iguanodon *to be found were its teeth (right). Their resemblance to those of living iguanas gave the dinosaur its name. Hadrosaurs had a battery of teeth for grinding food (below).*

bumps, while their cousins the ceratopsians evolved a bony frill that, in some later animals, extended well back over the shoulders. Pachycephalosaurs were all bipedal and were readily identified by their thick, bony heads, often ornamented with lumps and spikes of solid bone. They probably used their heads like battering rams against predators or to display for mates.

Ceratopsians were mostly quadrupedal with particularly large heads. The size of the head was further exaggerated by the neck frill and horns and spikes on the face.

Ornithischian plant-eaters dealt with food in different ways from their saurischian relatives. Some, such as the thyreophorans, probably snipped food into tiny pieces in the mouth using small, sharp teeth. Thyreophorans were unable to grind their food like later marginocephalians and ornithopods, which had batteries of teeth that formed grinding plates.

DIFFERENT SPECIES *of hadrosaurs ate different sorts of plants, so they were able to live together without taking one another's food supply.*

THE MEAT-EATERS

Meat-eating dinosaurs ranged dramatically in size. There were some no taller than a chicken and, at the other end of the scale, one that was the largest predator ever to walk the Earth.

The meat-eating dinosaurs are all within one group—the theropods. Some of the earliest known dinosaurs are thought to be theropods, giving them the longest history of any dinosaur group. In fact, because it is now established that birds evolved from theropods and should therefore be considered to be living theropods, this group has a continuous history that stretches over 230 million years.

Theropod fossils tend to be rarer than those of their plant-eating relatives, and they also display a higher diversity of types. Around 40 percent of all valid dinosaur genera recognized by paleontologists are theropods, but most of them are represented by only a single specimen. Because of this, we know less about the interrelationships of theropods than we do of other dinosaur groups, and the way they are organized into groups changes regularly as new material comes to light.

THEROPOD FEATURES

Most theropods were lightly built with large heads. They usually had bladelike teeth that often had serrated edges. Long, slender legs gave them greater speed than most other animals; they were all bipedal. Typically they had long, curved claws that tapered to spiked tips, especially on the hands. They also had hollow bones, a feature that would help birds, their descendants, take to the air and fly.

Another common feature of the theropods was some development of air-pockets or pneumaticity of the skull and vertebrae. They had at least five vertebrae connected to the hip, and an extra joint in the mandible that allowed the jaws to flex sideways to accommodate large pieces of food.

TYRANNOSAURUS (*left and above*) *had sharp, curved teeth—a typical feature of meat-eating dinosaurs.*

DIFFERENT THEROPODS *had various ways of handling their food.*
❶ *Compsognathus, a coelurosaur, caught and ate prey with its hands.*
❷ *Baryonyx used the huge, hooklike claw on its hand to spear its prey.*
❸ *Oviraptor cracked open eggs with its beak.*

TYPES OF THEROPODS

The theropods can be divided into two basic groups—the ceratosaurs and the tetanurans. The ceratosaurs are known almost exclusively from rocks of the late Triassic and early Jurassic, although some of the theropods from the late Cretaceous of Africa and South America may also be in this group. Typical of the ceratosaurs were dinosaurs such as *Dilophosaurus* and *Coelophysis*. They had four functional fingers on the hand, and clawed toes on the foot.

The tetanurans include all the other theropods. This group of dinosaurs had a maximum of three functional fingers and a foot with three large toes plus a smaller one on the inside of the foot.

COELOPHYSIS *was an agile predator that used its strong, clawed hands to grab small prey, such as the* Planocephalosaurus *that these two Coelophysis are chasing up a tree. These and some other meat-eaters also preyed on insects, such as cockroaches and dragonflies, frogs, mammal-like reptiles, and early mammals.*

wounds on the animals that they preyed on.

Recent discoveries, in particular those from China, reveal an abundance of small theropods closely related to birds that firmly establish the link between the two groups. Several of these dinosaurs, such as *Sinosauropteryx* and *Caudipteryx*, have been found preserved with feathers or featherlike structures in place around theropod skeletons. This raises the possibility that such features could have been much more widespread among the theropods, but the vagaries of fossilization have not pre-served these features in other known forms.

Theropods were the major carnivores of the day, and they undoubtedly had as much diversity in their habits as do the carnivores of today. The discovery of tens of *Coelophysis* together at Ghost Ranch in New Mexico, and many *Allosaurus* at the Cleveland-Lloyd Quarry in Utah, demonstrates that at least some theropods probably lived in groups of several animals. But the majority of theropod fossils have been isolated finds, which suggests that most theropods lived alone.

Carnosaurs were a group of (mostly) large tetanurans that included dinosaurs such as *Allosaurus* and *Sinoraptor*. This group has changed significant-ly in recent years with many former members being placed in other groups. Originally the carnosaurs were grouped together based solely on their large size, but former members such as *Tyranno-saurus*, the largest of them all, are now no longer thought to be carnosaurs.

Coelurosaurs were mostly Cretaceous tetanurans and included giants such as mighty *Tyrannosaurus* and all the dinosaurs most closely related to them.

More bizarre coelurosaurs included the slashing dromaeosaurs, the ostrichlike ornithomimosaurs, and the strange, crested oviraptors. Ornithomimosaurs and ovi-raptors were virtually tooth-less. This raises the question of what they fed on and how they managed to process their food. Dromaeosaurs had very birdlike skeletons as well as a deadly, retractable killing claw on the second toe that could inflict massive, slashing

THEROPOD DINOSAURS

had serrated teeth (right). Megalosaurus had new teeth ready to move into position to replace those that wore out (left).

Oligocene   36 million years ago

DINOSAUR MYTHS

The seemingly impenetrable mysteries that surrounded the dinosaurs and their ancient world acted as a powerful stimulus to the human imagination.

Even though modern science has dispelled many of the common myths about dinosaurs, these ancient animals continue to stimulate our imaginations.

Every culture has its myths of threatening monsters, sometimes depicted as fire-breathing dragons. It is hardly surprising, then, that early discoveries of huge, ancient bones fed the popular imagination with fantasies of everything from giant early humans to ancient elephant ancestors. Dragons are no longer in fashion, but our fascination with large reptiles has not diminished.

It has been suggested that the widespread human interest in reptiles is similar to the reaction that other primates, such as apes, display when they are confronted with snakes. They treat them warily. The fact that many medieval depictions of dragons are remarkably snakelike perhaps supports this point of view. On the other hand, many human cultures have coexisted quite comfortably with large reptiles, such as snakes or crocodiles.

Yet another argument suggests that early human encounters with fossils, perhaps by contact with trade caravans that crossed remote regions, helped to spawn fantastic legends of mythical beasts. There is no way of knowing.

One advantage of our fascination with dinosaurs is that it can inspire in young children an interest in science— imbuing them with a strong sense of the distant past and the processes by which organisms evolve and become extinct.

SCROTUM HUMANUM

When Robert Plot, a professor of "chymistry" at Oxford University, published the first description of a dinosaur bone from Cornwall, in England, in 1676, he had no idea that it was part of an extinct reptile. His description of it as an "enigmatic thighbone" indicates his bewilderment about its origins. In Plot's day, such fossils were generally interpreted as remains of giant humans or animals that had died in the biblical flood of the Old Testament. To suggest that it might have come from an extinct animal would have been anathema at that time—it would have implied that God had erred in creating unnecessary or imperfect creatures. Plot at first thought that the bone might have been part of an elephant brought to Britain by the Romans, but he concluded that it was the knee-end of the thighbone of a giant human. Scientists are now confident that Plot's bone was part of the thighbone of *Megalosaurus*.

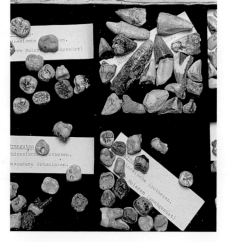

FOSSILS *from China (right) were traditionally sold as "dragons' teeth" and were supposed to have medicinal qualities. (Above right) Brookes' illustration of Scrotum humanum. (Top) A mythical dragon.*

A DEPICTION *from 1423 of the animals leaving Noah's Ark. Early finds of dinosaur fossils were thought to be the bones of giant animals that did not survive the biblical deluge.*

Another common myth is that dinosaurs had two brains. Many dinosaurs had enlarged neural canals within the hip region. Some people have speculated that, as the brain cavity seemed to be small, a mass of nerve tissue may have filled this hip cavity and helped to coordinate some of the functions a larger brain might otherwise have performed. Some living animals, however, also have enlarged neural canals in the hip region, and this debunks the myth of the two-brained dinosaur.

DEAD AS A DINOSAUR

This expression reflects several common misconceptions. The first is that dinosaurs were archaic, primitive, and ultimately unsuccessful animals. This ignores the fact that dinosaurs dominated the land for more than 150 million years and that throughout the Mesozoic they were highly diverse and lived alongside the supposedly superior mammals.

Thanks to evidence that has come to light in recent times, we now know that dinosaurs are, in fact, not extinct. The avian members of the group survive in their highly successful descendants—the more than 9,000 species of living birds.

MODERN BIRDS, *such as this kingfisher, are living descendants of the dinosaurs of the Mesozoic era.*

Although Plot's original bone is now lost, he illustrated it in a book, *The Natural History of Oxfordshire*, that he published in 1676 and it became a subject of considerable speculation. Almost a century later, in 1763, R. Brookes published an illustration of the bone, fancifully naming it *Scrotum humanum* because of its suggestive shape. The mythology of dinosaurs was well and truly launched.

INTELLIGENT DINOSAURS?

The common expression "dumb as a dinosaur" owes its origin to the observation that dinosaur brain cavities are small in relation to their body size. In fact, if we plot brain cavity volume against body mass, the brain size of dinosaurs falls within the predicted range for reptiles of similar size. Some of the smaller theropods have a brain–body mass ratio similar to that of living birds. Dinosaurs' brains seem so small only because their bodies are so large.

Notwithstanding their magnitude they must have been the bones of Men or Women.

The Natural History of Oxfordshire,
ROBERT PLOT (1640–1696),
English scientist

DINOSAURS IN ART, FILM, AND FICTION

Modern technology has provided the capacity to show dinosaurs very much as they existed throughout the Mesozoic era.

It is always a thrill to see dinosaurs mounted in a museum, but museum exhibitions do not give us a real sense of the living, breathing animals as they once were. It is unlikely that we will see cloned dinosaurs in the foreseeable future, so we are forced to use our imagination to bring them to life. A host of media has helped us to do this.

GRAPHIC ARTS

Soon after Buckland and Mantell published their discoveries of *Megalosaurus* and *Iguanodon* respectively in the early 1800s, artists began portraying them to a wide audience. Early renderings of *Iguanodon* look like giant iguanas, which was perfectly

MODELS *of King Kong and its pterosaur adversary were created for the classic 1933 movie (above right). (Right) An illustration from Conan Doyle's 1912 novel* The Lost World.

consistent with contemporary scientific opinion. The *Iguanodon* sculptures exhibited in London's Crystal Palace in 1854 were made under the supervision of Richard Owen and, although they look decidedly old-fashioned to modern eyes, they represented at the time the latest in scientific knowledge.

We can literally see the evolution of dinosaur science by watching artists' dinosaur depictions over the years. Most artists have worked with paleontologists to make their representations as accurate as

possible, given the latest scientific data.

In the late 1800s, dinosaurs were generally shown as agile, birdlike creatures, after Thomas Henry Huxley pointed out the similarity between the recently discovered *Archaeopteryx*, the oldest known bird, and other dinosaurs. By the early 1900s, however, the image of dinosaurs as sluggish, cold-blooded animals took over. It was not until the 1970s that energetic dinosaurs once again emerged in graphic depictions.

DINOSAURS IN FICTION

Novels that feature dinosaurs date back to the mid- to late 19th century. Most of the early examples featured explorers who stumbled upon prehistoric monsters in

WHAT IS IT?
HOW MUCH TERROR CAN YOU STAND?

GODZILLA vs. THE THING
IN COLORSCOPE

AN AMERICAN INTERNATIONAL PICTURE · Produced by TOHO CO., Ltd.

IN MOVIES *such as the Godzilla series of the 1950s, actors in costume played dinosaurs and other "monsters."*

DIGITAL REALISM

Modern films take advantage of computer animation technology. Many or all of the dinosaurs in *Jurassic Park* (inset) were completely digital. This made them more realistic than in any previous dinosaur movie. Movements looked more natural, and scenes combining dinosaurs and humans had a more convincing look. The dinosaurs themselves were also updated in line with contemporary scientific knowledge: Their tails were held off the ground and their gaits suggested elevated metabolic rates.

The recent television series from the British Broadcasting Corporation, "Walking with Dinosaurs," incorporates the latest scientific knowledge and represents dinosaurs, and the environment in which they lived, as accurately as is possible. The colors of the dinosaurs, of course, are entirely speculative.

sauropod who graced the world's screens from 1914. At that time, dinosaurs were usually "created" using models that were moved slightly between frame shots. This was the "stop-action animation" technique that worked so spectacularly with the *King Kong* sensation in 1933. Flat animation (cartoons) has always been popular and reached an artistic height with Disney's *Fantasia* in 1940. Many movies used human actors in costume to represent dinosaurs—a technique made famous by *Godzilla* (1956) and its sequels. Others resorted to putting fake horns or frills on living lizards or alligators and filming them up close.

THE 1940 MOVIE One Million Years *made numerous errors. For example, it brought the age of the dinosaurs much closer to the present and showed dinosaurs and humans existing together.*

remote jungle refuges—as, for example, in Sir Arthur Conan Doyle's *The Lost World* and some of Edgar Rice Burroughs's works. Sometimes—as in the movie version of Jules Verne's *Journey to the Center of the Earth*—the dinosaurs were deep underground. Early novels typically made use of extensive poetic license, often depicting dinosaurs and humans as existing together and also sometimes grossly exaggerating the size of the dinosaurs. But, to do them justice, most of these literary efforts drew on the scientific knowledge available at the time. As scientific concepts of dinosaurs evolved, the popular literature kept pace. Two recent novels by Michael Crighton —*Jurassic Park* and *The Lost World* (which bears no relation to Conan Doyle's book)—

depict dinosaurs much as paleontologists of the 1990s peceived them.

DINOSAURS ON SCREEN

For many of us, fictional and graphic representations are not enough. We want to see these fascinating animals in motion. Almost from the beginning of cinema history, dinosaurs featured in movies. One of the first movie cartoon stars was Gertie, a dancing

SOME MODERN DINOSAUR DETECTIVES

A second "golden age" of dinosaur discoveries that began in the 1970s continues unabated at the beginning of the 21st century.

What is often called the first golden age of dinosaur science stretched from the late 1800s to the early 1900s. Although scientists continued to study dinosaurs, research waned somewhat for 40 years after about 1930. In the 1970s, however, there was a great resurgence in interest. This was due in part to John Ostrom's ground-breaking comparisons between birds and small theropods such as *Deinonychus*, as well as new evidence that suggested dinosaurs may have been warm-blooded. The work of Robert Bakker, a student of Ostrom, helped to make popular a "new view" of dinosaurs during the 1970s and 1980s that depicted these ancient creatures, not as sluggish and lizard-like, but as agile, active, and birdlike.

SOUTHERN HEMISPHERE FINDS

Some of the most important discoveries in recent decades were made in

JOHN HORNER *is most famous for his 1978 discovery, in the Montana badlands, of nesting colonies of the hadrosaur Maiasaura.*

DONG ZHIMING *(above) is a leading Chinese paleontologist. (Left) Philip Currie, at right, and fellow researchers analyze dinosaur finds in the laboratory.*

Argentina. Here, rocks from the late Triassic—the period when the earliest dinosaurs were living—are exposed. Thanks largely to the work of José Bonaparte and other Argentinians, the existence of the primitive dinosaur *Herrerasaurus* was established. In the 1990s, another Argentinian, Fernando Novas, and the North American Paul Sereno found and described a much more complete *Herrerasaurus* specimen. This filled in many of the gaps in our knowledge of this early meat-eater. Novas's and Sereno's work also yielded *Eoraptor*, an even more primitive dinosaur than

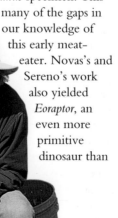

Herrerasaurus. These discoveries have provided valuable clues to just what the ancestral dinosaur may have looked like.

Other hotbeds of field work in the 1990s were sites in the former Gondwanan continents of South America, Africa, India, and Australia. One particularly successful project was in Madagascar, where parties led by David Krause and Catherine Forster, of the State University of New York at Stony Brook, excavated a late Cretaceous fauna, including armored sauropods and some bizarre theropods. These dinosaurs are, paradoxically, more similar to those found in the Cretaceous of South America and India and are related only remotely to Cretaceous dinosaurs from mainland Africa. This suggests that there was a dry-land connection between South America and India that was not joined to Africa.

PAUL SERENO, *of the University of Chicago, is seen here working in the torrid heat of the Moroccan Sahara to unearth the jaw of a Carcharodontosaurus.*

NORTHERN HEMISPHERE FINDS

In the Northern Hemisphere, paleontologists have been revisiting some of the classic localities that brought fame to earlier researchers. Philip Currie, of the Royal Tyrrell Museum of Paleontology in Alberta, Canada, has continued the tradition begun by Barnum Brown and the Sternberg family. His field work in the badlands of southern Alberta has brought to light new species as well as more complete remains of dinosaurs that had been found by earlier searchers.

Farther south, Jurassic and Cretaceous localities of the western United States continue to yield exciting material. These follow on from famous discoveries during the 1970s and 1980s by John Horner of Montana State University and his colleagues. One of their most celebrated discoveries was a set of duck-billed dinosaur nesting grounds at a place in the Montana badlands that Horner named "Egg Mountain." Horner called the new dinosaur *Maiasaura*.

If finding these fossils is so easy, why didn't anybody do it before?

It's a hard one to answer.

Digging Dinosaurs,
JOHN R. HORNER
(b. 1946–), American
paleontologist

NEW EMPHASIS

One important difference between the present dinosaur science and that of the previous golden age is the emphasis that paleontologists now place on small fossils. Earlier collectors did look for small dinosaurs, but they were more interested in finding giant skeletons that would make spectacular displays in major museums.

Paleontologists continue to collect large dinosaurs, but they now realize that small ones are just as important for helping us to understand the world of the dinosaurs. Most of the critical evidence that links birds to their dinosaur ancestry comes from smaller forms. Modern dinosaur scientists are also paying greater attention than their forebears did to the other animals—the mammals, lizards, crocodiles, turtles, and others—that lived alongside the dinosaurs.

These are good times for paleontologists. People who thought the days of great discoveries were over have been proved wrong. All over the world, fossils continue to be found—even at sites that were thought to have given up all their treasures as long as a century ago.

ROBERT BAKKER *is one modern researcher whose work has significantly expanded our knowledge of dinosaur biology. He is shown here with a reconstruction of a Stegosaurus skeleton.*

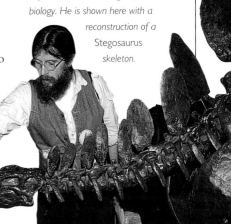

COLLECTING AND CARING FOR FOSSILS

Fossil-collectors, and the new finds they make, are essential to the health of the now-thriving science of dinosaur paleontology.

The first challenge that confronts any fossil-seeker is to know where to start looking. Not every spot is likely to yield dinosaur, or any other kind of fossils. Paleontologists, therefore, spend a good deal of time in the library, studying maps to find areas with an environment where rocks of the right age might be exposed. They then spend more time writing grant proposals to obtain necessary funds and permission to conduct their search. They also give careful thought to planning their excursion and deciding what materials and equipment—including tools, food, water, and fuel—they will need to take along.

IN THE FIELD

We often describe what paleontologists do as "digging" for dinosaurs. This is not entirely accurate, though some actual digging may well be involved. Once paleontologists have found likely fossil-bearing rocks, they spend most of their time walking round in a stooped, and eventually rather painful, posture with their eyes glued to the ground. They rarely stick their shovels into a random patch of earth and simply start digging. They let natural erosion start the process by exposing bones on hillsides. Most of the time they walk along valleys or washes, looking for fragments of bone that have been weathered from uphill. When bone is found, fragments are traced uphill until the source is located.

Searchers pinpoint the location of any fossil they find on a map. Modern Geographic Positioning System (GPS) technology has made this task a good deal easier than in past days. Having located their

BETWEEN 1909 AND 1912, *a huge German expedition excavated fossils at Tendaguru. Local workers did the digging and carried the fossils from the site to the nearest seaport (below). (Left) A collection of tools used for uncovering and extracting bones from rock.*

A DIGGING TEAM *in Niger (above) carefully cleans the exposed sections of the front and back limbs of a giant sauropod dinosaur. (Above right) Before removing a large and delicate fossil from the site, this team encloses it in a "jacket" of plaster and sacking and then reinforces the covering with wooden battens.*

fossils, paleontologists must then decide how to proceed. If the bones are small—no longer than an inch or two—they may simply be rolled up in toilet paper and taken back to camp; larger fossils will require quarrying and wrapping (a process described on pages 98–9).

Bringing large fossils out of the field can be a challenge. German-led field parties to Tendaguru, in present-day Tanzania, before World War I employed long convoys of local foot couriers, who carried the fossils more than 700 miles (435 km) to the port of Lindi. Fossils from western North America were hauled by mules in the late 19th and early 20th centuries. Today, we can use helicopters if funds permit. If not, we drag the fossils to nearby trucks.

FOSSILS IN MUSEUMS

When the fossils arrive at a museum, specially trained staff remove them from their surrounding matrix and then carefully clean them (this process is described on pages 100–1). The fossils must then be stored under special protective conditions. It is not a matter of simply putting them on shelves. Frequent changes in temperature or humidity can damage some specimens, and insects can destroy specimen labels. Fossils are kept in closed cabinets, which in many museums have sophisticated climate-control systems.

Fossils belong in museums, because that is where they are most readily accessible to the scientific community. While most fossils in museums are never displayed to the public, they make up the data set that paleontologists rely on for their research and for the advancement of their knowledge. The fossils that are on exhibit may well be the largest or the most attractive or dramatic, but they may not, from a scientific point of view, be the most significant. The fossils hidden away from public view often provide the basic sustenance on which the science of paleontology relies for its survival.

ESSENTIAL TOOLS

The first thing that any fossil-seeker, amateur or professional, needs to bring to the field is specialized knowledge and training. No one should try to quarry large vertebrate fossils without training in the proper methods, as poor quarrying can result in the destruction of critical information. Some museums and universities provide training programs for amateur fossil-hunters.

The next requirement is an understanding of the geography and geology of the region that is being investigated. A fossil collected without detailed information of where it came from will be of little value.

Yet another essential is a good supply of patience. For every exciting moment of a signficant find, there will inevitably be long, frustrating periods of fruitless searching.

Necessary tools include a rock hammer, a hand lens or magnifying glass, a camera, a map, a notebook and pencil, a set of dental picks or probes, a shovel (for some areas), and toilet paper and plastic bags, for wrapping and storing what you find. Perhaps the most essential single item, however, is a comfortable, good-quality, sturdy pair of working boots.

THE COLLECTOR'S CODE

FEDERAL LAW PROHIBITS DISTURBANCE OR REMOVAL OF FOSSILS OR OTHER MATERIALS ON MONUMENT LAND

Responsible fossil-collectors abide by a set of general rules that protect both important fossils and the rights of individual enthusiasts.

Collecting fossils can be a wonderful hobby, and paleontology is one of the few sciences where hard-working amateurs continue to make significant contributions—professional scientists cannot do the job alone. But fossils are a finite resource. How can we ensure that everyone benefits?

Most collectors understand the importance of not clearing a site of all its fossils. To do this would be to deprive other collectors of the chance to find anything—at least until further erosion exposes more fossils, which can take a long time. Respect for the landscape you are working in is another essential feature of the collector's code. Try to minimize the effect you have on the site you are working— do not dig more than is necessary, return flipped rocks to their original position, and take all garbage out with you.

GREAT CARE *needs to be taken when hunting for fossils in dangerous situations, such as on steep slopes. Dangerous animals can also be a source of concern.*

SAFETY

Safety, of course, is a prime consideration, especially as field geology has the highest mortality rate of all the sciences. Most fatalities involve automobiles, so be particularly careful at road cuts. If there are dangerous or poisonous plants and animals in the region, learn how to identify them. Dress sensibly for the weather conditions; in bright sun wear long-sleeved shirts, polarized sunglasses, and apply sunscreen. Drink plenty of water—in a desert you should consume at least

CHILDREN *can be taught the "rules" of responsible fossil-hunting from an early age (left). For the protection of valuable fossils, signs such as this (top left) must always be heeded.*

4 pints (2 L) a day. Take care around steep slopes, and do not collect fossils during a thunderstorm. Such safety advice is basically common sense, but there can often be a temptation to abandon common sense in the excitement of the search.

FOSSILS AND THE LAW

The laws governing the collection and possession of fossils vary from country to country, and sometimes within different regions of the same country. It is important that you familiarize yourself with the laws that apply by contacing a local state, provincial, or federal government office. If you wish to prospect on private property, it will always be necessary to obtain permission from the landowners and to make sure that both parties have a clear understanding of where the collecting will occur, what can be collected, and what will become of it. In some places, collecting on public lands is unregulated; in others, there are strict laws that place limits on, or even prohibit, collecting.

There is considerable debate about which fossils should be owned by amateur or commercial collectors. While most people would agree that mollusks or sharks' teeth should be readily available for school use, few would sanction the private sale of, say, a *Tyrannosaurus* skeleton. Sometimes, though, the line between what is acceptable and what is unacceptable is not so clear-cut. For guidance on such matters, you might get to know members of local

THE BEST PLACE FOR A FOSSIL

Some professional paleontologists argue that all important fossils should be kept in museums, where they benefit the community at large. Some collectors who sell fossils would rather let the market decide where fossils should go. Caught in the middle are the amateurs, who represent the majority of those who collect fossils.

There are many practical reasons for important fossils to be kept in museums. For one, according to the rules of biological nomenclature, any specimen that forms the basis of a new species must be in a museum. Fossils in private collections cannot be used to designate new species. For another, private collections are not permanent. When a private collector dies, the collection may well be sold, dispersed, or discarded. For example, one of the seven known skeletons of *Archaeopteryx*—the so-called Maxburg specimen—was in a private collection. When, during the 1980s, the owner died, the specimen disappeared. This is much less likely to happen in a museum. Even if a museum closes, its collection will almost certainly be transferred to another museum, and there will be a public announcement of the transfer.

amateur fossil clubs and any professional paleontologists who are working in your area. They will be able to advise whether a particular fossil is important enough to warrant becoming part of a museum collection.

If you find something that clearly belongs in a museum—such as a bone from any vertebrate—or if a fossil is so fragile that it is likely to be damaged unless professionally handled, record its location as precisely as possible. Take a photograph of the fossil in the ground, and contact a nearby museum.

THIS FOSSIL COLLECTOR, *excavating the hip bone of a* Tarbosaurus, *is well protected against the rigors of the Mongolian climate.*

WHERE TO LOOK AND WHAT TO LOOK FOR

A bit of homework, and asking the right people, can make dinosaur hunting that much easier and, hopefully, rewarding.

Most large museums with dinosaur fossils in their collections have on staff one or more vertebrate paleontologists. An important part of their job is to find more specimens for study and display. Where should they—or, indeed, amateur fossil hunters—start to look? We know that only certain types of rocks contain dinosaur fossils, and a little preliminary research will help pinpoint some promising places to start looking.

A SUNSET SCENE *in the Pariah badlands in Utah, USA. Areas such as this can provide rich pickings for fossil-hunters.*

OLD AND NEW SITES

The museum itself is a good starting point. It will have records of where its existing dinosaur fossils came from and may well be willing to share this information with serious enthusiasts. You could try going back to the original site—if one fossil was found there, more will probably be waiting there. Every season erodes the rock a little more, and specimens that were not visible on one trip may come to light on a later one. If the dinosaurs at your museum were collected a long time ago, it's definitely worth a return visit—a good deal of erosion will have occurred in the meantime.

Some rock types, or formations, tend to produce many more fossils than others, so if you find out which

LOOKING FOR FOSSILS DOWN UNDER

Australia, like North America, has many of the features that make for good dinosaur hunting—large areas of Mesozoic sedimentary rocks and extensive arid zones. So far, however, only a handful of dinosaur fossils have been found in Australia. Why is this? Part of the answer lies in Australia's small population and the relatively few paleontologists who live there. Probably more important, though, is the fact that during the Ice Ages there were no ice sheets covering Australia— there were no glaciers to produce fresh and rapid erosion of the rock, as there were in the United States. By contrast, the exposed rocks in Australia have been weathering away relatively slowly for millions of years. We know that there were lots of dinosaurs living in Australia, because there are many good trackways—dinsosaur footprints. Dinosaur hunting, Australian style, can involve much more tracking than digging.

THIS LATE *Cretaceous bone bed in Dinosaur Provincial Park in Alberta, Canada, contains remains of the horned herbivore Centrosaurus.*

The museums that are most successful at finding new specimens are often those that work closely with amateur fossil hunters and local landholders. Many amateur fossil hunters have great enthusiasm for the task and are very skilled. Local landholders know their properties intimately and are more likely than anyone else to find exposed fossils there. Some museums run special programs to help amateur fossil hunters and have also found it worthwhile to spend time showing landholders what to keep an eye out for.

If you think you have found a dinosaur fossil, do not start digging or removing bones. If you have picked up a bone that you think might be a fossil, examine it closely and then put it back exactly where you found it. Look for other pieces that may be lying close by but do not disturb the site unnecessarily. When you have finished searching, take whatever you have found back to the museum for identification. Always note the exact position it was in and ensure that you will be able to find the site when you return.

formations produced the existing specimens, you can use a geology map to establish where else you might find these formations. The map will also show places where creeks or roads cut through the rock—these are good places to look.

Often small shards of bone are weathered from a freshly exposed fossil and are moved downslope by wind or water. If you see small pieces lying loose, follow the slope or the creek back to see where they may have come from.

IN THE FIELD

Look for sites where the rock is exposed at the surface. Rocky deserts are good fossil-hunting areas. Dry creek beds, cliff faces, road cuttings, and even quarries are also useful places to look.

GANTHEAUME POINT, *near the town of Broome, on the north-west coast of Western Australia, is thought to be the site of ancient dinosaur mud baths. Many well-preserved fossils have been found in the region.*

EXCAVATING IN THE FIELD

A dinosaur dig is like a crime scene—great care is needed

to get every last clue out of the ground.

When a dinosaur fossil has been found, it needs to be removed from its site as quickly as possible. Even though it may have been lying in the rock for hundreds of millions of years, as soon as it is exposed to the elements, it can begin to erode away more rapidly. However, the removal must be done with great care. Unlike the search, which can be undertaken by a small group, or even a single person, the dig can require a large amount of equipment and a lot of people. Just organizing a dig team can itself be a major task.

FINDING THE WHOLE FOSSIL

Once a part of the fossil has been uncovered, the rest of it needs to be found as soon as possible—whether it is lying in the ground nearby or has been washed halfway down a gully. If the fossil has only just been exposed and has not been moved by rain or wind, then it is likely that the rest of it will be nearby, so the surrounding rock needs to be carefully cleaned away. Gentle brushing and chiseling is usually the best way to do this. In the meantime, someone with sharp eyesight can be set to work combing the surrounding area, or even farther afield, in search of other pieces that may have been dislodged, or washed or blown downhill.

EXCAVATING *a Tyrannosaurus hip bone in South Dakota, USA.*

MAPPING

When you are confident that most of your dinosaur fossil has been found, you can start mapping and labeling the bones. The information gained from the position each bone was found lying in can be as important as recovering the bone itself. Therefore, no bone should be removed until its position has been mapped on a grid, and the bone's orientation and number have been marked.

Usually the site is marked out with string and pegs. Then a 1 square meter (11 sq. ft) frame supporting a 10-centimeter (4-in) grid is placed in turn over each square meter of the site. A map of the bones lying within each section of the square is drawn onto graph paper and sections of the grid are photographed progressively.

LABELING

In order to create an accurate record of the original find, you need to photograph the bones in the context of the surrounding site before beginning to remove them— a combination of stills and

PALEONTOLOGISTS *work carefully to excavate a well-preserved giant marsupial skeleton at Cox's Creek, New South Wales, Australia.*

video is best for this. Making an annotated sketch of your find can also be useful.

The next step is to label the bones. Stick-on labels are not satisfactory, as they are liable to become detached and get lost. Each bone should, therefore, be labeled with permanent marker. If you feel you need to consolidate a fragile bone with PVA glue before removing it, you should apply this before you start labeling, as the glue will usually smudge the ink.

REMOVING FOSSILS

Once you have mapped, labeled, and photographed the fossils, you are ready to remove them. How to do this depends on both the nature of the fossils and the substrate they are lying in. If fossils are lying loose in sand or dirt, you can simply lift them out with care and then wrap and pack them. Very small fossils, such as bone fragments or teeth, can be wrapped in tissue paper and stored for the time being in small boxes.

Larger bones will need to be wrapped temporarily in cloth or hessian and placed in straw in large crates.

If fossils are embedded in solid rock—or even hard clay—small pieces may be very carefully chiseled out, but you will usually have to remove some of the surrounding rock as well. You can cut the rock around the fossil or you can work existing cracks in the rock to make a series of blocks containing the fossil bones. A wide range of tools is useful for this, depending on the hardness of the rock. Chisels, picks, rock saws, hammers, pneumatic drills, and even humble penknives can be brought into play. Be careful, that an over-enthusiastic blow with a hammer does not send an unwanted crack running through the middle of a bone.

PROTECTING FOSSILS FOR TRANSPORTATION

When the block is ready to come out, you need to protect the fossil on the upper surface—and sometimes the sides of the block as well—before moving it. If the block and fossil are fairly strong, they may only need to be protected by a wrapping of thick, wet newspaper. If, however, the fossil and the block are more fragile, you will have to gradually build up layers of tissue or toilet paper, newspaper, cloth, and then plaster, before it is ready to move. In all, it can be a very painstaking and time-consuming process.

After all that lengthy, delicate work, the moment of taking a mighty blow to separate the block from the rest of the rock can be very satisfying! The block is then lifted, the underside is wrapped—and, if necessary, plastered in the same way as the top—the outside of the block is labeled, and your fossil is ready at last to be transported back to the museum. Once there, it will be prepared for closer study and, perhaps, ultimately, to go on exhibition.

A SCIENTIST *employs a grid to map out the site of a fossil find in Shell, Wyoming, USA.*

BACK IN THE LABORATORY

The long job of preparing a new specimen will require the expertise of some of the museum's most specialized staff.

When a fossil arrives at the laboratory, it is unpacked as soon as possible, while details of the dig are still fresh in people's minds. The bones are then cleaned and washed and, if necessary, they are laid out in the way that they were found in the field (a sandpit can be very useful for heavy blocks). Piecing together a large dinosaur can be like doing a heavy three-dimensional jigsaw puzzle with an unknown number of pieces and no picture to work from!

A TECHNICIAN *in Colorado, USA, uses a powerful magnifying glass as she prepares the fused vertebrae of an Allosaurus specimen.*

The specimen must now be conserved and prepared for study and display. There are many types of conservation treatments available for fossils, and each fossil has different conservation requirements. Many bones are treated with glues to strengthen them and most are kept at a constant level of temperature and humidity.

PREPARING THE FOSSIL
Preparing the fossil involves removing any matrix (rock) that is still attached to the bone. Exactly what type of preparation, and how much is undertaken, depends on the

SECTIONING

Many fossils preserve the details of the dinosaur's internal anatomy in fine detail. Fossilized dinosaur bone often contains a great deal of information about the microscopic structure—or the "histology"—of the bone. This can be studied by examining a very thin slice of the bone under a microscope. Dinosaur bone histology has been used to investigate rates of growth and even to examine the question of whether dinosaurs were "warm-blooded."

Sometimes skulls can be sectioned (cut down the middle) to reveal the internal structure of the skull bones, such as the braincase, which is not visible from the outside of the fossil. Sectioning usually involves

simply cutting the skull in half, but in some cases the fossil has to be finely sliced to reveal the internal detail. This is called serial sectioning. The problem with this process is that it destroys the fossil, so casts and photographs need to be taken throughout. When the preparation is finished, the cast is the only record of the fossil that remains.

Recently, powerful medical scanners, which use X-ray imaging to "slice" the specimen, have been employed in serial sectioning. This has the advantage of revealing the internal detail of the specimen without destroying it, but it does have one important drawback—often the X-ray is unable to distinguish between the rock and the bone.

FRENCH PALEONTOLOGIST

Professor Jean Lautier is seen here in his laboratory in Paris, France, restoring a fossilized dinosaur egg. The preparation and study of dinosaur eggs and embryos have greatly enhanced our understanding of dinosaur reproduction and nurturing behavior.

nature of the fossil and the matrix and what sort of study or display is eventually required.

In most cases this preparation will be either mechanical or chemical. Mechanical preparation involves the use of special drills and chisels to remove the matrix from the bone. With chemical preparation, a weak acid is used to remove the rock. Each method has its advantages and its problems.

Mechanical preparation, if it is not undertaken very carefully, can seriously damage the fossil. A variety of tools—such as small rock saws, chisels, and pneumatic percussion tools—may be used. The advantage of mechanical preparation is that it does not involve the use of chemicals that could eventually harm the specimen.

Acid preparation can be used when the phosphatic bone is enclosed in a carbonate rock, such as limestone—a weak acid will react with the matrix but will not affect the bone. Acid

preparation can expose delicate structures that would be damaged if they were subjected to mechanical preparation, but there is always the danger that the acid will penetrate into the specimen and slowly attack it from within.

The preparation of a dinosaur fossil requires a technician with a great deal of skill and patience. He or she must work closely with a paleontologist who has a detailed knowledge of the dinosaur's anatomy. As the bone is slowly

CHINESE PALEONTOLOGIST

Xi Xing carefully brushes and blows sediment from the pubis of a recently discovered therizinosaur, Beipiaosaurus. This dinosaur, at about 7 feet (2.2 m) long, is the longest therizinosaur yet found. This fossil shows evidence that Beipiaosaurus had protofeathers.

revealed, observations are noted and photographs are taken. In order to establish how much rock must be removed, the bone is compared with those of other known dinosaurs.

Preparation can be very time-consuming—a specimen that takes a week to excavate may take years to prepare.

BRINGING A NEW DISCOVERY TO LIFE

The first stage in turning fossil into animal is to slowly and painstakingly reconstruct the dinosaur's anatomy

The skeleton of an animal is made up of its bones, teeth, cartilaginous structures, and skeletal muscles. A dinosaur fossil is usually a record of some part of the animal's skeleton. Because they contain many hard minerals, the bones and teeth of dinosaurs are the structures that most often get fossilized.

A lot can happen to a bone during the millions of years the fossil spends lying in the ground—it may get broken, squashed flat, or crushed. A paleontologist needs to allow for any distortion that may have occurred as a result of one or more of these occurrences. With complex structures such as skulls, this can be quite tricky. Sometimes, powerful computer programs can be used to "uncrush" such misadventures as a badly distorted specimen. In many instances, however, a specimen that is badly crushed or eroded may well prove too difficult to study, or to put on display.

CONSTRUCTING THE SKELETON

A fossil specimen rarely preserves more than a fraction of the 200 or so bones that are present in the vertebrate body. The study or display of the specimen, then, usually involves the reconstructing of the entire skeleton. Scientists therefore try to fill in the missing bones. If, for example, the tail from one specimen is missing, it may still be possible to reconstruct it by making comparisons with the tail of a similar dinosaur that has been found. This process assumes that the restorers have a sound knowledge of anatomy and are familiar with many

EXPERT DRAWINGS *of dinosaur fossils are of great importance in helping scientists reconstruct the appearance of individual dinosaurs.*

different types of dinosaurs. It can work only if specimens of similar dinosaurs have already been found.

Once researchers have reconstructed the bones, they then need to fill in details of ligament and muscle attachments. They can do this by examining the bones for particular clues that indicate at which points ligaments or muscles were attached to them. Using the muscles of living animals as a guide, they can then make a generalized reconstruction of the animal's muscle system.

RECONSTRUCTIONS *of Baryonyx have been made using only some bones (highlighted here) found in a clay pit in south-eastern England.*

SCIENTISTS AT BOEING'S ROCKETDYNE LAB *at Rocket County, California, use sophisticated X-ray techniques to examine a recently discovered Tyrannosaurus skull.*

OUTSIDE AND INSIDE

By this stage of the process, we have a good idea of the size and shape of the dinosaur. The details of the outer covering—the skin, scales, and feathers—are not often preserved in fossils. If a specimen has no fossil record of these features, there is little point in trying to reconstruct them. General reconstructions of dinosaur scales and feathers are based upon those rare fossils that do preserve them, as well as on comparisons with living species.

The internal organs of dinosaurs have only rarely been preserved. Almost all reconstructions of the organ system of dinosaurs are based on comparisons with living animals rather than on direct fossil evidence. However, gut contents are sometimes preserved, and these can provide much valuable information about a dinosaur's last meal—or whether it swallowed grit or stones to help it process its food. There is no fossil evidence that can help us to discover what color dinsoaurs were when they were alive.

GETTING IT WRONG

Educated guesswork often plays an important role in bringing dinosaur fossils to life. However, attempts to reconstruct dinosaurs from limited fossil evidence can lead to false conclusions.

During the 1930s, for example, a pair of huge arms, each with a set of ferocious claws, was uncovered in Mongolia. The dinosaur from which they came was named *Deinocheirus*, or "terrible hand," and for decades there was intense speculation about this mystery dinosaur, which was assumed to be a monster carnivore.

The truth turned out to be otherwise. A series of finds in the late 1980s revealed that the huge arms belonged to an unusual group of theropod dinosaurs—the therizinosaurs. They had long necks, small heads, and teeth that seemed suitable for eating plants. The hips were huge, but the legs and tail were quite short. We now think *Deinocheirus* was a dinosaurian version of the extinct giant ground sloth, which sat on its haunches and used its powerful arms and claws to rip vegetation and pull it toward the mouth.

BARYONYX *reconstructions have been made from the basic skeleton by using information about the muscles, skin, and internal organs of modern reptiles.*

LEARNING FROM LIVING CREATURES

Living animals provide us with many valuable leads in our attempts to reconstruct the appearance, anatomy, and behavior of dinosaurs.

The next stage in bringing a dinosaur to life is to reconstruct its behavior and how it lived, using living animals as a model. Because dinosaurs are so long extinct, "bringing them to life" requires some lessons from living animals. Because the entire process of reconstructing a fossil species relies heavily upon a knowledge of the anatomy, behavior, ecology, and evolution of living creatures, a good paleontologist must be a biologist as well as a geologist.

COMING TO CONCLUSIONS

All the characteristics that make up a living animal are influenced both by the features it inherited from its distant ancestors and by the way it lives in its present environment. We can see from living animals that some features vary greatly within one species, while others are surprisingly consistent within a group of related animals. Despite this, making careful and informed comparisons with living species

can allow scientists to reconstruct, with a fair degree of confidence, features that cannot be seen in the fossil.

Though we can observe different scale patterns in living reptiles, species of the same type of reptiles tend to have similar scale patterns. The scale type in dinosaur skin was, therefore, able to be reconstructed by looking at the scale of the living archosaurs—crocodilians and birds. The assumption that dinosaur scales were like those of these living archosaurs has indeed been confirmed by the presence of dinosaur skin in some rare, well-preserved dinosaur fossils.

Because there are plenty of fossilized dinosaur eggs in existence, it may seem reasonable to conclude that all dinosaurs laid eggs. This, however, is based on an absence of evidence to the contrary. Nevertheless, a comparison

THE POWERFUL JAWS
and the teeth of present-day crocodiles are reminiscent of those of their reptilian ancestors, including many carnivorous dinosaurs.

with living animals shows that all living archosaurs—including birds, alligators, and crocodiles—also lay eggs. This suggests that archosaurs are somehow constrained to laying eggs and does back up the notion that all dinosaurs were egg-layers.

EVIDENCE FROM TEETH

The diet of living animals is strongly related to tooth and jaw structure, and it is reasonable to think, for example, that dinosaurs with teeth like those of modern carnivores tended to eat meat. Most reconstructions of dinosaur diets are based on comparisons with the tooth shapes of living animals. The similarity between the teeth of iguanas and those of *Iguanodon*—which inspired that dinosaur's name—indicates that, like the iguana, *Iguanodon* ate plants. By comparing the wear marks on herbivorous dinosaur teeth with those found in living animals, we can even tell how

MODERN BIRDS *(above) are descended from dinosaurs like Archaeopteryx (left), but have lost the claws on the wings, the teeth, and the long bony tail of their distant relatives. The feet, however, are similar.*

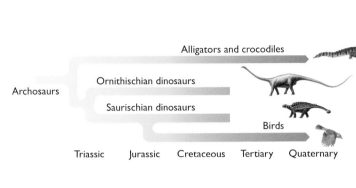

	Triassic	Jurassic	Cretaceous	Tertiary	Quaternary

Archosaurs — Alligators and crocodiles — Ornithischian dinosaurs — Saurischian dinosaurs — Birds

dinosaurs ate their plant diet. Again, similarities between the teeth of modern Komodo dragons and those of tyrannosaurs suggest that, like Komodo dragons, tyrannosaurs trapped rotting pieces of meat in their teeth and used this rotten meat to inflict infectious bites on prey, although there is no evidence to support this theory.

Teeth, though, are not always a reliable indicator of what an animal eats. Modern bears, for example, have a typical "carnivorous" tooth structure, but many eat large amounts of plant matter.

Unlikely Comparisons

Sometimes useful comparisons can be made between the most unlikely animals. The fact that dinosaurs laid eggs means that large dinosaurs must have

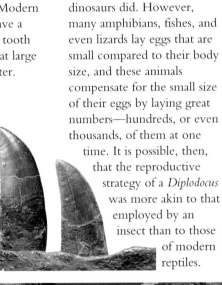

BIRDS ARE THE LIVING ANCESTORS *of the dinosaurs that became extinct at the end of the Cretaceous period. Alligators and crocodiles are not directly descended from dinosaurs, but they share a common ancestry through the archosaurs.*

laid eggs that were tiny in comparison to their body size. This is because the mechanical properties of eggshell place an upper limit on the size of eggs. No large animal alive today lays such relatively small eggs as dinosaurs did. However, many amphibians, fishes, and even lizards lay eggs that are small compared to their body size, and these animals compensate for the small size of their eggs by laying great numbers—hundreds, or even thousands, of them at one time. It is possible, then, that the reproductive strategy of a *Diplodocus* was more akin to that employed by an insect than to those of modern reptiles.

The Color Question

Comparisons with living creatures can tell us nothing about the colors of dinosaurs. Among living animals, there can be enormous color variation between closely related species, or even between individuals of the same species. Even though there are some general rules of thumb about an animal's coloration, there are so many exceptions that it is impossible to know without direct evidence what an animal's color might have been. All sorts of colors and patterns—including patterns for camouflage, mimicry, threat, or display—can be imagined. In dinosaur reconstruction, the color is often left for the artist to decide. Esthetic appeal rather than scientific rigor is, in this case, the only useful criterion.

DISCOVERIES OF FOSSIL TEETH *from dinosaurs such as Tyrannosaurus (above left) have helped to establish evolutionary links between ancient and modern reptiles. The mud nest (left), which contained Oviraptor eggs, was discovered in 1922. These were the first dinosaur eggs ever discovered.*

SCIENTISTS' CHANGING PERCEPTIONS OF DINOSAURS

Changing attitudes and beliefs, as well as new scientific discoveries, have radically altered people's attitudes toward dinosaurs.

Our understanding of dinosaurs is an ever-changing collage of thousands of individual pieces of work, spanning more than a century and a half.

Scientific research is always performed in the context of a society's prevailing philosophies and prejudices, and the study of dinosaurs is no exception. When researchers first began to take a serious scientific interest in dinosaurs, more than 150 years ago, their understanding of the world, and of the place of humans in it, was very different from the picture we have today.

EARLY IDEAS ABOUT DINOSAURS

The first scientific studies of dinosaur fossils were undertaken in the early 19th century, long before Charles Darwin published his theory of evolution in 1859.

Sir Richard Owen, the famous British scientist who in 1842 first coined the word "dinosaur" was certainly no believer in evolution. Even after Darwin's theory had been published, he remained

EARLY DEPICTIONS *of Iguanodon showed it with a sharp horn on its snout. By the early 20th century artists were getting closer to the truth (left). Richard Owen (below left) described many new species of ancient animals.*

a firm opponent of evolution. Indeed, he even opposed the "progressive" idea, which gained currency during the Victorian era, that there had been a progression over time from simple to more complex life forms. Owen actually saw dinosaurs as a more advanced form of present-day reptiles.

At the time that Owen was formulating these ideas, during the 1840s, the only evidence at his disposal consisted of a few bones and small pieces. No complete skeleton had yet been found, and very few dinosaurs had yet been named. From this meager evidence, Owen concluded that all dinosaurs were large, four-footed animals, rather like modern rhinoceroses. He had no concept at this stage of the

fleet-footed, smaller dinosaurs, such as *Struthiomimus*, nor, of course, of a dinosaur such as the birdlike *Archaeopteryx*, which evolutionists would later hail as the perfect example of the "missing links" predicted by Darwin. When, in the 1860s, soon after it was discovered, Owen did study a specimen of *Archaeopteryx*, not surprisingly he declared it to be a bird, and not a dinosaur at all.

SLUGGISH BEHEMOTHS

Victorian ideas of progress through time led many 19th-century scientists to view dinosaurs as some kind of failed experiment—and a cursory examination of the fossil record does suggest that they were replaced by far more "advanced" animals, the mammals. But subsequent discoveries and a growing body of research data have revealed that stereotypes of dinosaurs as sluggish behemoths are both facile and inaccurate. We now

THE DINOSAUR–BIRD LINK

Proving that birds evolved from dinosaurs is considered by some biologists to be the most important achievement of dinosaur paleontology. The birdlike nature of some dinosaurs was first noticed by the English anatomist T.H. Huxley in the 1860s, and the discovery of *Archaeopteryx*, with its mix of dinosaur and avian features, was quickly recognized by Huxley as strong evidence for a dinosaur–bird link. In the early part of the 20th century, however, the idea that birds were descended from dinosaurs fell out of vogue. It was not revived until 1969, when John Ostrom described *Deinonychus*, noting its many birdlike features. Since then a wealth of "dino-birds" have been discovered, especially from China, and the theory that birds are "living dinosaurs" is now widely accepted. This belief has in turn radically altered ideas about dinosaurs generally. Modern reconstructions of many dinosaurs show them as warm-blooded, feathered, active animals—just like their descendants.

have a keen appreciation of the considerable complexity and uniqueness of dinosaurs and of their remarkable success in holding their position as the dominant land animals on Earth for more than 150 million years.

MODERN RESEARCH

As new evidence comes to light, new perceptions of dinosaurs will emerge to challenge our present views. There is still a great deal to learn—as evidenced, for example, by the present scientific debate about whether dinosaurs were warm- or cold-blooded animals, and by the uncertainty that still surrounds their extinction. The discovery of a single new specimen may well radically alter our thinking. It may show some anatomical feature that has not been seen before in a particular family of dinosaurs, or it may reveal a new species of a genus of dinosaurs that are already known from other specimens. It could even be the first specimen of a kind of dinosaur that has never been seen before.

No one would suggest for a moment that the final word has yet been been said about any species of dinosaur. At the end of the 21st century many of our present ideas about dinosaurs will no doubt seem as quaint and archaic as the prevailing ideas put forward by scientists and dinosaur enthusiasts 100 years ago seem to us today.

ON NEW YEAR'S EVE *in 1853, a formal dinner was held at London's Crystal Palace inside a model of* Iguanodon. *This illustration shows Richard Owen, at the head of the table, and of* Iguanodon, *proposing a toast.*

IN THE MUSEUM

Teaching the public about dinosaurs is one of the most important and interesting jobs in the museum.

A visit to a natural history museum can provide the stimulus for an enduring interest in dinosaurs. Every year millions of people go to museums to marvel at the earthly remains of these wonderful animals. There are literally hundreds of dinosaur displays worldwide. Each of them is the end result of a huge amount of work and may involve a very substantial financial outlay.

MOUNTING AN EXHIBITION

Big museums often employ large teams of highly skilled designers and artists whose job it is to put the exhibitions together. The process starts with a paleontologist and a designer deciding on the basic theme of the display. They may, for example, want to demonstrate some aspect of a new theory about a species or highlight a recently acquired specimen. They must decide whether they are going to use mounted skeletons, full-sized "life" reconstructions, robotic dinosaurs, or computer graphics—or perhaps some combination of all of these.

If a skeleton is to be mounted for display, a cast needs to be made of each bone. Fossilized bone is usually too heavy to mount safely in a display, and fossil material that is part of an exhibit is not readily available for further study. Casts are often nearly impossible to tell from the original specimen, and they have the advantage that they can be touched by the visitors. Such hands-on experiences are an important feature of many displays. Making a cast of every bone in a dinosaur specimen, as well as sculpting the missing bones, can take many months. Constructing a full-sized model of what the dinosaur looked like in life is often quicker and less expensive. Either way, the job is long and painstaking.

Many recent dinosaur displays use robotics to make the animals move as if they were alive, or use "interactive" computer graphics to add to the effect. Others depict the fossil skeleton as it lay in the ground. This involves re-creating the original dig site. Sometimes a fossil is prepared

A PALEONTOLOGIST *(above) makes casts of dinosaur fossils. (Right) Artist Brian Cooley, in his studio in Calgary, Alberta, Canada, adds feathers to a model of a juvenile* Sinosauropteryx.

LARGE MOUNTED DINOSAUR SKELETONS *dominate the central hall of the Oxford University Museum of Natural History in Oxford, UK.*

in a special laboratory that can be viewed by the public. This means that the fossil is on display almost as soon as it arrives at the museum.

MUSEUM COLLECTIONS

Once a specimen has been displayed to the public, it is sent to its final resting place—as part of the collections of a museum. Though the displays are the most visible part of any museum, the collections are actually more important—they provide a safe keeping place where the fossils can be made available to be studied by paleontologists.

Each specimen is first catalogued, so that it can be located easily, and then stored in whatever way is considered most appropriate for that particular fossil. Usually the collections are kept at a constant temperature and humidity. This lessens the risk of pyrite disease, a condition

that can occur when iron pyrite, a mineral found in many fossils, begins to oxidize. They are also kept reasonably free of dust.

Specimens that represent new species are called "types," and these must be stored with extra care. In many cases, these types will never be allowed to leave the museum. Collections of some very large museums can be so extensive that no one is really sure exactly what they contain. The best place to find a new

It is always the final thing that one must research with the most perseverance.

Letter to Gideon Mantell,
BARON GEORGES CUVIER
(1769–1832), French anatomist

species of dinosaur is sometimes among old bones lying on a museum shelf!

THINKING AGAIN

Some mounted dinosaur skeletons have been in place as long as the museums that house them. However, as ideas about dinosaurs can change over time, older displays can look quite dated. The *Diplodocus* skeleton in the main gallery at London's Natural History Museum has been there for many years. When it was first mounted, scientists believed that dinosaurs used to drag their tails on the ground. The mount, therefore, had its tail snaking along behind it. When it was realized that all dinosaurs held their tails well clear of the ground so that the tail counterbalanced the front half of the body, it was necessary to remount the tail. Visitors can now walk underneath the huge tail and gaze up at a newer, livelier *Diplodocus*.

USING MUSEUMS AND FINDING MORE INFORMATION

A range of facilities, from the local natural history museum to the World Wide Web, are valuable sources of information and advice about dinosaurs.

Your local natural history museum is undoubtedly the best place to go for information about dinosaurs and the science and practice of paleontology. The staff of most museums are more than happy to have interested members of the public volunteer their services. In fact, most museums need, and actively seek, volunteers to help with such activities as specimen preparation, cataloging, and exhibit interpretation. Often, at a museum display, it is the well-informed volunteer, rather than the professional staff member, who will be the most readily accessible person who can provide guidance to visitors and answer their questions.

A TRICERATOPS *skull at the Museum of the Rockies in Montana, USA. The first complete skull was discovered in Wyoming in 1889 by John Bell Hatcher.*

MUSEUM SERVICES

Many museums offer guided field trips to collect fossils. Some of these excursions are to nearby localities and give people an opportunity to increase their appreciation of the fossil resources of their own neighborhood. Others are field excursions to places farther afield, and while they usually cost more money, they can be of great value in broadening the perspectives of an amateur enthusiast.

Some museums have extensive training programs for interested people. These programs include, among other things, training in the techniques of fossil collection and preparation. In certain instances, these courses can even lead to the earning of an official certificate. Sometimes, there are outreach programs that establish contact between particular museums and fossil collectors in their region. Such programs are mutually advantageous to the museum and the collectors. The museum gains access to material it would not have obtained using its ownlimited resources, and the collectors are able to make use of professional expertise in fossil preparation and preservation, as well as in data management.

Advances in robotics have enabled some museums to build realistic, lifesized models of dinosaurs that move just like the real thing. They can even roar and grunt. These displays are always a great attraction in museums, especially with children.

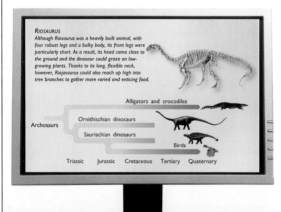

RIOSAURUS
Although Riosaurus was a heavily built animal, with four robust legs and a bulky body, its front legs were particularly short. As a result, its head came close to the ground and the dinosaur could graze on low-growing plants. Thanks to its long, flexible neck, however, Riojasaurus could also reach up high into tree branches to gather more varied and enticing food.

Alligators and crocodiles

Archosaurs
Ornithischian dinosaurs
Saurischian dinosaurs
Birds

Triassic   Jurassic   Cretaceous   Tertiary   Quaternary

DINOSAUR WEB SITES, *especially those listed on the opposite page, can be a valuable source of information.*

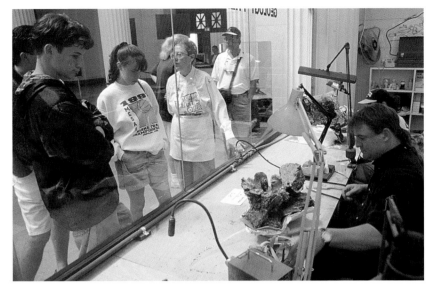

VISITORS *to the Field Museum in Chicago, USA, have the opportunity to watch museum technicians preparing fossils for display. Here they observe a technician working on part of the Tyrannosaurus skeleton known as "Sue." Sue went on display in May 2000.*

CLUBS AND SOCIETIES

In many areas there are local fossil-collecting clubs or societies. Apart from offering the chance to socialize with people with similar interests, these organizations give enthusiasts opportunities to increase their knowledge and expertise by swapping field stories, discussing fossil identification, and learning about new collecting sites. Clubs frequently sponsor field trips to known sites and often have connections with a number of local museums.

Some professional paleontological societies permit serious amateurs to become members. These societies, however, are not for the casual dilettante. Members of these societies receive subscriptions to journals, some of them highly technical, which bring them up-to-date with the latest research in paleontological matters. They can also attend annual general meetings, where they can rub shoulders with prominent paleontologists and benefit from their knowledge and expertise. A list of the major international societies is given at the end of this page.

BROWSING THE WEB

Until relatively recently, the easiest way of obtaining information about dinosaurs was by consulting books in a library or by making contact with other enthusiasts. Today, the World Wide Web provides everyone with a rich source of contact information.

There are, however, some traps to be aware of. Many fossil-related sites are put together by people with little or no paleontological expertise and can be mainly sources of misinformation——they can even be wilfully misleading at times. However, if used judiciously, the Web can be used to reach out to other collectors and to find societies or clubs you might want to join.

Following are the major international paleontological organizations. Their links pages can point enthusiasts toward a wealth of reliable information about matters relating to dinosaurs.

Society of Vertebrate Paleontology
www.vertpaleo.org/
Paleontological Society
www.paleosoc.org/
Palaeontological Association
www.palass.org/
Paläontologische Gesselschaft
www.palaeontologische-gesellschaft.de/palges/
The Dinosaur Society
www.dinosaursociety.com/

THIS RECONSTRUCTED SKELETON *of the giant theropod Allosaurus is on display at the Carnegie Museum in Pittsburgh, USA*

USING THE GUIDE TO DINOSAURS

This guide gives detailed information about a wide range of dinosaurs, some whose names are very familiar, others that have only recently been discovered.

Fossil finds have provided vital clues about the environments that dinosaurs inhabited, as well as what these creatures looked like, the food they ate, and how they lived. The text in the following pages, supported by lifelike illustrations, presents the most up-to-date facts and figures about almost 80 different species of dinosaur.

The taxonomic data *in the panel at the top of a page indicates the group and family to which a dinosaur belonged. This provides a quick reference to how dinosaurs were related to each other.*

The genus name *of each dinosaur is followed by a guide to how it is pronounced and on which syllables the emphasis should be placed.*

Field Notes Panels

■ *Meaning of the dinosaur's name*

🦖 *Approximate maximum size*

■ *Period in which it lived*

■ *Where it lived*

🏛 *Museums around the world where fossil remains or reconstructions of the dinosaur are on display*

Ceratopsia/Psittacosauridae

Psittacosaurus
sih-TAK-oh-saw-rus

FIELD NOTES

■ Parrot lizard

🦖 4 feet (1.25 m)

■ Early Cretaceous

■ Mongolia; China; southern Siberia; Thailand

🏛 American Museum of Natural History, New York, USA; Paleontological Museum, Moscow, Russia; Academy of Science, Ulan Baatar, Mongolia; Department of Mineral Resources, Bangkok, Thailand

Psittacosaurus was discovered in Outer Mongolia in 1922, in the early stages of the famous expeditions undertaken by the American Museum of Natural History between 1922 and 1925. Henry Osborn named it for the beaklike appearance of its face. It is known from a number of well-preserved skeletons, which represent about eight different species from Mongolia, southern Siberia, and northern China, as well as from some lower jaw fragments that were discovered in northern Thailand.

Psittacosaurus was one of the earliest dinosaurs to show the typical beaked face of the ceratopsian group. This beak, which was supported by a single median bone—called the "rostral bone"—is the one feature that distinguishes the ceratopsians from all other dinosaur groups.

Psittacosaurus was one of the smallest and most primitive of 194 the ceratopsians. It lacked the

Psittacosaurus lived in the early Cretaceous of Mongolia and other parts of Asia about 90 million years ago. It had relatively long forelimbs and large, grasping hands.

well-developed frill and horns that were typical of more advanced ceratopsians, yet, along with the hard keratinous beak, it had the characteristic skull shape of a ceratopsian. It also featured, in common with later ceratopsians, the high palate and the sharp, slicing teeth with self-sharpening edges that were well suited to nipping off and shredding hard plant matter.

Psittacosaurus's hindlimbs were longer—although only slightly—than its forelimbs, which suggests that it could have moved about in an upright position for short distances. It may have done so to avoid attacks from predators or to forage in low-hanging tree branches.

Some skeletons of *Psittacosaurus* contain fossils of gastroliths—stomach stones that helped the animal to break down plant matter inside its stomach.

One of th[e] neocera[...] *Protocera[...]* developed frill t[...] from the face a[...] However, it la[...] the more adva[...] the group, alth[...] featured a sma[...] that may have [...] tinous horn, [...] modern rhin[...]

Protocerat[...] in the late C[...] Mongolia. [...] dinosaur w[...] fossils foun[...] of Natural [...] that Roy [...] 1922 and [...] fossils has [...] *Protocerat[...]* in herds.

The [...] eggs and [...] were th[...] the dis[...] famous [...] interlo[...]

The main text *gives a detailed word portrait of each dinosaur, its appearance, and lifestyle.*

The illustrated banding *at the top of the page is a visual pointer to indicate that the page is part of the Guide to Dinosaurs.*

The main illustration *provides a vivid recreation, based on the latest scientific research, of how the animal would have looked in life.*

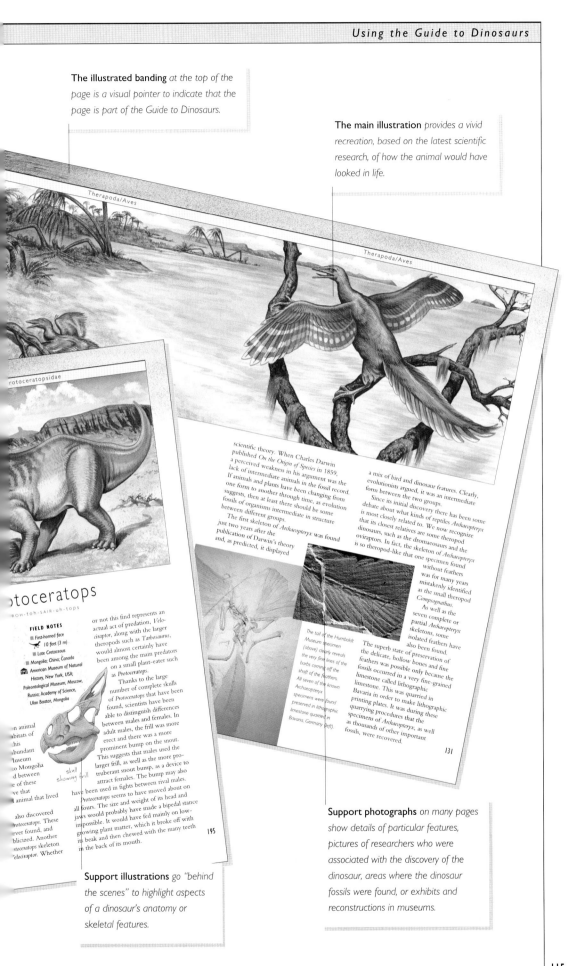

Therapoda/Aves

Therapoda/Aves

rotoceratopsidae

scientific theory. When Charles Darwin published *On the Origin of Species* in 1859, a perceived weakness in his argument was the lack of intermediate animals in the fossil record. If animals and plants have been changing from one form to another through time, as evolution suggests, then at least there should be some fossils of organisms intermediate in structure between different groups.

The first skeleton of *Archaeopteryx* was found just two years after the publication of Darwin's theory and, as predicted, it displayed

a mix of bird and dinosaur features. Clearly, evolutionists argued, it was an intermediate form between the two groups.

Since its initial discovery there has been some debate about what kinds of reptiles *Archaeopteryx* is most closely related to. We now recognize that its closest relatives are some theropod dinosaurs, such as the dromaeosaurs and the oviraptors. In fact, the skeleton of *Archaeopteryx* is so theropod-like that one specimen found without feathers was for many years mistakenly identified as the small theropod *Compsognathus*.

As well as the seven complete or partial *Archaeopteryx* skeletons, some isolated feathers have also been found.

The superb state of preservation of the delicate, hollow bones and fine feathers was possible only because the fossils occurred in a very fine-grained limestone called lithographic limestone. This was quarried in Bavaria in order to make lithographic printing plates. It was during these quarrying procedures that the specimens of *Archaeopteryx*, as well as thousands of other important fossils, were recovered.

The tail of the Humboldt Museum specimen (above) clearly reveals the very fine lines of the barbs coming off the shaft of the feathers. All seven of the known Archaeopteryx specimens were found preserved in lithographic limestone quarried in Bavaria, Germany (left).

131

otoceratops

ROH-toh-SAIR-uh-tops

FIELD NOTES

First-horned face
10 feet (3 m)
Late Cretaceous
Mongolia; China; Canada
American Museum of Natural History, New York, USA; Paleontological Museum, Moscow, Russia; Academy of Science, Ulan Baator, Mongolia

skull showing frill

or not this find represents an actual act of predation, *Velociraptor*, along with the larger theropods such as *Tarbosaurus*, would almost certainly have been among the main predators on a small plant-eater such as *Protoceratops*.

Thanks to the large number of complete skulls of *Protoceratops* that have been found, scientists have been able to distinguish differences between males and females. In adult males, the frill was more erect and there was a more prominent bump on the snout. This suggests that males used the larger frill, as well as the more protuberant snout bump, as a device to attract females. The bump may also have been used in fights between rival males.

Protoceratops seems to have moved about on all fours. The size and weight of its head and jaws would probably have made a bipedal stance impossible. It would have fed mainly on low-growing plant matter, which it broke off with its beak and then chewed with the many teeth in the back of its mouth.

195

Support illustrations *go "behind the scenes" to highlight aspects of a dinosaur's anatomy or skeletal features.*

Support photographs *on many pages show details of particular features, pictures of researchers who were associated with the discovery of the dinosaur, areas where the dinosaur fossils were found, or exhibits and reconstructions in museums.*

Eoraptor

EE-oh-RAP-tuh

Perhaps the most primitive of all known dinosaurs, *Eoraptor* was found in late Triassic deposits—in the harsh badlands of the Ischigualasto Basin in north-western Argentina—in 1993. It was discovered by Paul Sereno, Fernando Novas, and their colleagues in the same set of deposits as *Herrerasaurus*, another relatively primitive dinosaur. *Eoraptor* was discovered almost accidentally—one member of the team was on the verge of discarding a rock before noticing that it contained teeth. A closer examination revealed that this rock contained a complete skull. This led to further examination of the site and the unearthing of a complete skeleton of *Eoraptor*, a hitherto unknown dinosaur almost 230 million years old.

From the evidence of this skeleton, we can be fairly confident that *Eoraptor* was a bipedal predator and a very primitive relative of the theropods. It ran mainly on its hindlimbs but may have walked on all fours from time to time. Although it had five fingers on its hand, the

FIELD NOTES

■ Dawn thief

🦖 3 feet 3 inches (1 m)

■ Late Triassic

■ North-western Argentina

🏛 Not on display

fifth finger was very small. Like other theropods, *Eoraptor* had thin-walled, hollow bones in its arms and legs and stood on feet dominated by the three middle toes. However, unlike in other theropods, the first toe may have helped to support *Eoraptor* when it was walking.

Eoraptor's serrated teeth indicate that it was a meat-eater and the grasping hands on the end of the forelimbs suggest that it was capable of handling prey almost as large as itself. Although we cannot accurately reconstruct this animal's predatory behavior, it seems to have been a very active, fast-moving hunter that preyed upon a range of lizard-sized animals, including some warm-blooded ancestors of today's mammals.

This X-ray photograph shows Eoraptor's skull being held by the hand of Paul Sereno, the paleontologist whose team discovered this dinosaur in Argentina.

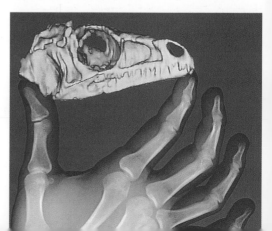

Herrerasaurus

huh-RARE-uh-SAW-rus

FIELD NOTES

- Herrera's reptile
- 6 feet 6 inches (2 m)
- Late Triassic
- North-western Argentina
- Instituto Miguel Lillo, Tucumàn, Argentina; Field Museum, Chicago, USA

Named after Victorino Herrera, a goat farmer who in 1963 found the skeleton of this dinosaur in the Ischigualasto Basin in north-western Argentina, *Herrerasaurus* is one of the oldest of known dinosaurs. It was not until 1988, however, when North American Paul Sereno and Argentinian Fernando Novas, and their team of paleontologists, found a more complete skeleton of *Herrerasaurus*, that this dinosaur was finally described.

hindlimb and foot bones

Sereno's and Novas's description showed that *Herrerasaurus* was a very primitive theropod. As with other theropods, its bones were hollow and thin-walled, its teeth were serrated, and both its upper and lower jaws had a joint that allowed the mouth to open widely while the dinosaur was feeding.

Herrerasaurus's hand had long, sickle-shaped claws and an opposable thumb, which made this hand a formidable weapon for attacking and grabbing prey. Such a hand foreshadowed those of later theropods. However, unlike the hands of other theropods (except that of *Eoraptor*),

Herrerasaurus's hand had five fingers. The hand differed from *Eoraptor*'s in that the fifth finger was reduced to a single bone (the metacarpal) and was probably covered in tissue. Although the third finger—and not the second as in other theropods—was the longest, all of the first three fingers were very long. Slightly more advanced theropods lost the fifth finger entirely, and in later theropods the fourth finger also disappeared, producing the three-fingered hand we see in today's birds. We can catch a glimpse of the process of evolution by comparing the hands of *Eoraptor*, *Herrerasaurus*, and *Coelophysis* with those of later theropods.

Like *Eoraptor*, *Herrerasaurus* was probably a swift-moving, bipedal predator. As *Herrerasaurus* was about twice the size of *Eoraptor*, it may well have included its smaller contemporary in its prey. Larger predators no doubt hunted both *Herrerasaurus* and *Eoraptor*. These early dinosaurs occupied a small, though significant, niche in the world of 230 million years ago, representing about one-twentieth of all animals alive at that time.

117

Coelophysis

SEE-loe-FIE-sis

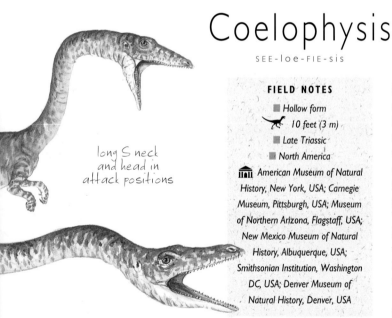

long S neck
and head in
attack positions

FIELD NOTES

■ Hollow form

🦖 10 feet (3 m)

■ Late Triassic

■ North America

🏛 American Museum of Natural History, New York, USA; Carnegie Museum, Pittsburgh, USA; Museum of Northern Arizona, Flagstaff, USA; New Mexico Museum of Natural History, Albuquerque, USA; Smithsonian Institution, Washington DC, USA; Denver Museum of Natural History, Denver, USA

from Ghost Ranch represented adult animals only about 5 feet (1.5 m) long. The remains of some of the smallest specimens were found inside the rib cages of larger *Coelophysis*. This led to a theory that these dinosaurs gave birth to live young. What now seems most likely, however, is that *Coelophysis* was a cannibalistic animal that included members of its own species in its diet.

Most theropod hands had three fingers. The most primitive known theropods, however, had more than three. *Coelophysis* had four fingers on each hand, but the fourth digit was extremely small and may have been invisible in a living animal.

We cannot be certain why so many skeletons turned up in a fairly restricted area at Ghost Ranch. However, geologists think it likely that this region, like tropical regions today, was prone to seasonal wet and dry cycles. It is possible that a whole herd of *Coelophysis* died during a particularly harsh dry season and was buried rapidly during a flood at the beginning of the ensuing wet season.

When Edward Drinker Cope described *Coelophysis* in 1881, his description was based on only a few fragments he found in late Triassic sediments in New Mexico. Thanks, however, to a treasure trove of more than 100 skeletons of various sizes that a team of paleontologists discovered in 1947 at Ghost Ranch in northern New Mexico, *Coelophysis* is now one of the best known of all nonavian dinosaurs. It is also one of the most primitive of known theropods.

Like most other basal theropods, *Coelophysis* was relatively small. Many of the specimens

118

Plateosaurus

PLAY-tee-oh-SAW-rus

Many skeletons of *Plateosaurus*—which is the best known of the prosauropods—have been found in the south of Germany, as well as in France and Switzerland. A harmless plant-eater with numerous small, pointy teeth, all of uniform size and shape, *Plateosaurus* roamed in herds around the northern part of Laurasia during the late Triassic. It was one of the first of the large dinosaurs.

Plateosaurus had stout limbs that supported the considerable weight of the animal as it walked on all four legs. As with other prosauropods, *Plateosaurus*'s hind legs were stronger than the front ones and were able to take the weight of the creature when it reared up, either to reach higher branches for food, or possibly to defend itself against attack—a particularly large claw on each thumb would have been an effective weapon in such a circumstance. Its long tail would have acted as a counterbalance to the long, thick neck.

FIELD NOTES

■ Flat lizard

🦖 26 feet (8 m)

■ Triassic

■ Western Europe

🏛 American Museum of Natural History, New York, USA; Senckenberg Nature Museum, Frankfurt, Germany; State Museum of Natural History, Stuttgart, Germany

Small ridges of bone around *Plateosaurus*'s mouth supported cheek pouches that could have held a mouthful of leaves while the front teeth went on stripping more leaves from a tree branch. The cheeks also served to prevent food from falling out of the mouth while the animal chewed, breaking the plant matter down into a mush that it could then swallow.

In at least one instance—near the German town of Trossingen—a herd of *Plateosaurus* appear to have been killed in a flash flood. The floodwaters buried the skeletons together in a jumbled mass. From this remarkable deposit paleontologists have been able to reconstruct what a herd of *Plateosaurus* probably looked like. It would have included animals of different ages—from very young to very old—that found safety in numbers as they moved about in search of food.

Paleontologists once thought that prosauropods such as *Plateosaurus* were the ancestors to those later giants of the Jurassic, the sauropods. Now, however, they are recognized as a separate group in their own right.

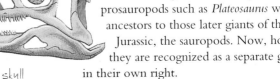

skull

119

Riojasaurus

ree-OH-juh-SAW-rus

Although *Riojasaurus* was a heavily built animal, with four robust legs and a bulky body, its front legs were particularly short. As a result, its head came close to the ground and the dinosaur could graze on low-growing plants. Thanks to its long, flexible neck, however, *Riojasaurus* could also reach up high into tree branches to gather more varied and enticing food.

Some paleontologists believe that sauropods such as *Riojasaurus* evolved their long necks in order to reach the ever-dwindling supplies of food as the world dried out at the end of the Triassic. As feed plants at ground level became scarce, plant-eaters that could reach up into the trees would have had a distinct advantage—and the further they could reach up, the greater that advantage would be.

FIELD NOTES

- La Rioja lizard
- 33 feet (10 m),
- Late Triassic
- La Rioja, Argentina
- San Miguel de Tucumán Museum, Argentina

Early finds of *Riojasaurus* were headless, and sharp, pointed teeth were often found with the skeletons. This led many to conclude that these dinosaurs were meat-eaters. We now know, however, that they were plant-eaters of elephantine proportions. The pointed teeth probably belonged to meat-eating dinosaurs that fed on the freshly dead carcasses, shedding teeth as they feasted.

Dinosaurs such as *Riojasaurus*, which had a small head at the end of a long neck, needed to minimize the weight of the neck so that they could hold the head high without a great effort. *Riojasaurus* was one of the first long-necked dinosaurs with hollow bones in its neck.

Unlike other prosauropods, *Riojasaurus* and its close relatives were probably not capable of rearing up on their back legs. The sheer weight of these very solid animals kept them firmly anchored on all four legs.

Members of this group, which included *Roccosaurus* and *Vulcanodon*, lived as far afield as South America, South Africa, and England. In the late Triassic, there was no Atlantic Ocean to separate the landmasses.

complete skeleton

Ceratosaurus

seh-RAT-oh-SAW-rus

detail showing bony head and snout

FIELD NOTES

- Horned lizard
- 20 feet (6 m)
- Late Jurassic
- North America
- American Museum of Natural History, New York, USA; Dinosaur National Monument, Jensen, Utah, USA; Smithsonian Institution, Washington DC, USA

Ceratosaurus gets its name from the blunt horn at the end of its snout. It is the best known of several therapods that had such a horn. *Ceratosaurus* also had two other short, thin horns, one over each eye.

We do not know for certain what function these three horns served. They may have worked as social signaling devices, allowing individual *Ceratosaurus* to recognize each other across the floodplains on which they lived. They may also have distinguished male and female animals, but as we have only a few specimens of this dinosaur, we do not know if the sexes had differently shaped horns. The horns do not seem large enough to have played a role in defense.

Ceratosaurus is known mainly from the late Jurassic Morrison Formation of the western United States. The Morrison Formation is famous for its incredible diversity of sauropods (including *Apatosaurus* and *Brachiosaurus*) as well as for the more common theropod *Allosaurus*. Teeth gathered from the Tendaguru Formation of Tanzania, which is roughly the same age as the Morrison, may represent a very large species of *Ceratosaurus*. This, however, has not yet been firmly established.

Throughout most of the 20th century, paleontologists grouped all large theropod dinosaurs, including *Ceratosaurus*, under the heading "carnosaurs." Smaller theropods were classed as "coelurosaurs." Beginning in the 1980s, however, scientists began to pay much greater attention to evolutionary relationships when classifying dinosaurs. Scientists now think that *Ceratosaurus* was only distantly related to most other large theropods, such as *Allosaurus* and *Tyrannosaurus*. It was probably more closely related to the smaller-bodied coelophysoids and to the Abelisauridae (including *Abelisaurus*)—an intriguing group that lived throughout the Southern Hemisphere during the Cretaceous period. *Ceratosaurus* certainly demonstrates the danger in classifying dinosaurs solely on the basis of size, as large body size evolved numerous times within the theropods.

121

Dilophosaurus

die-LOH-foh-SAW-rus

ossils of *Dilophosaurus*, one of the earliest of the large theropods, were first found in 1942, in early Jurassic sediments in Arizona. More recently, remains of this animal have also been discovered in China. It was a close relative of *Coelophysis*, and, like this earlier dinosaur, had four fingers on each hand. The fourth finger, however, was very small and probably had no function.

A pair of paper-thin crests ran along the top of *Dilophosaurus*'s snout, projecting behind the eyes. It is these crests that gave the dinosaur its name. Other coelophysoids, such as *Syntarsus*, a contemporary of *Dilophosaurus*, had similar crests, but those of *Dilophosaurus* were larger and more extensive. We do not know what function these crests may have served, but we can tell that they were very delicate structures that had shallow pits and holes which may have served as air sacs. These pits are more apparent in the Chinese specimen, but they are also clearly noticeable in the specimens from Arizona. It is possible that *Dilophosaurus* utilized its crests as a signaling device—to distinguish one species from another. Alternatively, it may have used them as a way of telling males and females apart.

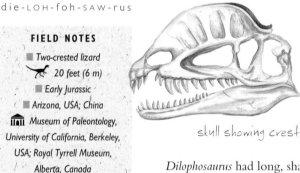

FIELD NOTES

■ Two-crested lizard

🦕 20 feet (6 m)

■ Early Jurassic

■ Arizona, USA; China

🏛 Museum of Paleontology, University of California, Berkeley, USA; Royal Tyrrell Museum, Alberta, Canada

skull showing crest

Dilophosaurus had long, sharp, pointed teeth. However, it probably did not use its teeth to kill its victims; some scientists have suggested that its jaws were not strong enough to enable it to fed on live prey. It probably used its clawed hands and feet to kill its victims and then fed on their carcasses. It may also have scavenged animals killed by other predators.

In the movie *Jurassic Park*, *Dilophosaurus* is depicted with an extendable frill—rather like that of an Australian frillnecked lizard—and also as spitting poison. However, imagination has ruled the day here for there is no evidence for either of these features. The idea that it spat venom may have resulted from suggestions that, as it seemed unable to attack live animals with its teeth and jaws, it killed them with poison. However, as no living crocodylian or bird is known to use venom in this way, there can be no reason to suppose that *Dilophosaurus* did.

122

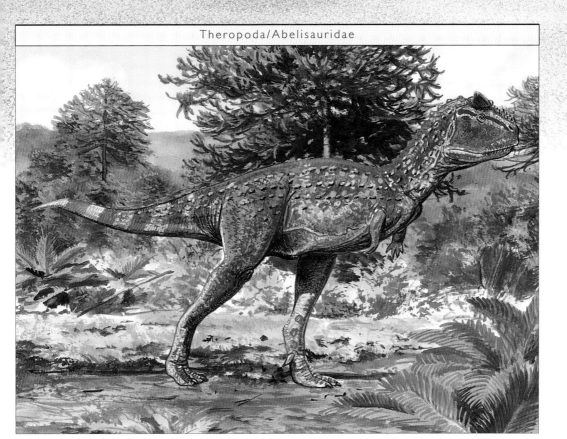

Carnotaurus

KAR-noh-TAW-rus

This bizarre-looking thero-pod, distinguished by a pair of sharp, stout horns that projected outward above its tiny eye sockets, is known from only a single specimen, discovered by the Argentinian paleontologist José Bonaparte in the Patagonia region of Argentina. The skeleton was almost complete, and there were also some impressions of skin. The whole specimen was protected by a large concretion—a section of very hard rock. These skin impressions, which covered much of the body and part of the skull, had a "pebbly" texture and were described as being reptilelike. However, the scales on the skin did not overlap as they do on some reptiles.

Bonaparte concluded that *Carnotaurus* belonged to a hitherto unknown theropod family—the Abelisauridae. Other specimens of abelisaurids—probable relatives of the Jurassic ceratosaurids—were later discovered in Argentina, India, and Madagascar. This provides us with evidence that these landmasses were connected at some point during the Jurassic or Cretaceous, as no positively identified abelisaurids have been found anywhere else, including mainland Africa.

FIELD NOTES

■ Meat bull

🦎 16 feet (5 m)

■ Early Cretaceous

■ Argentina

🏛 Los Angeles County Museum, USA; Argentine Museum of Natural Sciences, Buenos Aires, Argentina

Like all other abelisaurids, *Carnotaurus* had sharp, serrated teeth that seemed to splay out from the sides, giving the face a rather triangular look when seen from the front. One feature in which *Carnotaurus* resembled the tyrannosaurids was in its forelimbs, which seem ridiculously short for an animal of its size. But the construction of the forelimb was different from that of the tyrannosaurids. In tyrannosaurids, the bones in the lower limb (the radius and the ulna) were shorter than the bone in the upper limb (the humerus), but they were still substantial bones, and the hand had only two functional fingers. In *Carnotaurus*, the radius and ulna were so short that they looked almost like wrist bones rather than lower arm bones, and the hand had four digits.

skull showing lower jaw and teeth

123

Allosaurus

AL-oh-SAW-rus

FIELD NOTES

■ Other lizard

🦖 39 feet (12 m)

■ Late Jurassic

■ Western United States

🏛 American Museum of Natural History, New York, USA; Smithsonian Institution, Washington DC, USA; Cleveland Museum of Natural History, USA; Los Angeles County Museum, USA; Denver Museum of Natural History, USA; Utah Museum of Natural History, Salt Lake City, USA; Dinosaur National Monument, Jensen, Utah, USA; University of Wyoming Geological Museum, Laramie, USA; Science Museum of Minnesota, St. Paul, USA

Abundant remains of *Allo-saurus* have been found in the late Jurassic Morrison Formation of the western United States. In many ways this was the quintessential large theropod. It had a deep skull and jaws filled with flattened, serrated teeth. The jaws were capable of bending outward in the middle, thus enlarging the mouth. The head was perched on a slender but strong neck. The dinosaur's forelimbs were heavily muscled and ended in powerful, three-fingered grasping hands with enormous claws. Its hind-limbs were massively constructed to support the animal's weight but were proportioned for rapid movement. *Allosaurus* stood on only the middle three of its five toes. This formidable predator was probably one of the major threats to any ornithopods, steg-osaurs, or sauropods that lived near it. An earlier name for *Allosaurus*—*Antrodemus*, or the "nightmare dragon"—reflects its predatory dominance.

Most specimens of *Allosaurus* are less than 26 feet (8 m) long, but some larger ones, including "Big Al," which is on display at the University of Wyoming, are close to 40 feet (12 m)—and scientists think that Big Al was still immature when it died.

Allosaurus belonged to the allosauroids, a group that peaked in the late Jurassic and early Cretaceous and was found in most parts of the world. This group includes *Giganotosaurus*, *Carcharodontosaurus*, and *Sinraptor*. By the late Cretaceous, however, other large theropods such as the tyrannosaurids and abelisaurids had replaced the allosauroids in most places.

An Allosaurus *and* Camptosaurus *battle in this reconstructed "death scene" from 140 million years ago.*

Eustreptospondylus

yoo-STREP-toh-SPON-dee-lus

FIELD NOTES

■ Well-curved vertebra

🦖 23 feet (7 m)

■ Middle Jurassic

■ Oxfordshire, Buckinghamshire, UK

🏛 Oxford University Museum, Oxford, UK

One of the larger predators of its time, *Eustreptospondylus* was a powerfully built ceratosaurian that displayed the usual theropod pattern of long, birdlike hindlimbs and quite short arms. The skull was moderately deep—unlike those of more lightly built ceratosaurians such as *Dilophosaurus* and *Coelophysis*—but it lacked the prominent horns of the larger, and later, *Ceratosaurus*. Although not closely related and somewhat smaller, *Eustreptospondylus* was similar in its proportions to the more widely known carnosaur *Allosaurus*.

A number of herbivorous dinosaurs are known from the same habitat, including early kinds of ankylosaurs (*Sarcolestes*), stegosaurs (*Lexovisaurus*), and the sauropods *Cetiosaurus* and *Cetiosauriscus*. *Eustreptospondylus* presumably preyed on plant-eaters smaller than itself, such as the ankylosaurs and stegosaurs, but it may also have attacked the larger cetiosaurs.

Eustreptospondylus is known mostly from a single fragmented, but well-preserved, skull and skeleton that were discovered in the mid-19th century in Oxfordshire, England.

Some bones that normally fuse in adult theropods, such as the postfrontal and postorbital in the skull, and the sacral vertebrae, remain unfused in this individual, so it is thought to have been immature when it died. For many years, this was the best-known of the large carnivorous dinosaurs from the whole of Europe.

Before many complete dinosaur skeletons were known and paleontologists began to realize their sheer diversity, nearly all the remains of large Jurassic carnivores in Europe were assumed to belong to *Megalosaurus*, which was one of the first dinosaurs to be named. It was more than a century—not until 1964—before Dr. Alick Walker recognized inconsistencies with other megalosaur material and the "*Megalosaurus*" skeleton from Oxford was finally given a new name, *Eustreptospondylus*. As well as this skeleton, some limb bones from another site are included in the same species.

skull showing weight-saving holes

125

Megalosaurus

MEG-uh-loh-SAW-rus

Despite its very familiar name and its association with the early scientific study of dinosaurs, we know surprisingly little about *Megalosaurus*. William Buckland's description of *Megalosaurus* in the early 1820s was the first formal description of a nonavian dinosaur. It was based on a collection of fossil fragments—including parts of a leg, a shoulder, a hip, and a jaw—that gave only scant clues to the appearance of the living animal. Since then, remnants of a wide variety of large theropods— including *Tyrannosaurus* and *Allosaurus*—have been mistakenly identified as belonging to *Megalosaurus*, and there are probably still some misidentifications waiting to be corrected.

The available evidence—which relies largely on fossils of the jaws, teeth, and hip bones—suggests that *Megalosaurus* was a massive animal. Estimates of its length

FIELD NOTES

■ Great lizard

🦕 30 feet (9 m)

■ Middle Jurassic

■ England

🏛 Natural History Museum, London, UK

saw-edged teeth

and height vary, though it probably grew 30 feet (9 m) long, stood up to 10 feet (3 m) high, and weighed about 1 ton (1.02 t). Like other theropods, it was probably a bipedal predator with a grasping, three-fingered hand. However, even this is still speculative, as no remnants of *Megalosaurus*'s forelimbs have yet been found. Huge, inward-pointing footprints found in trackways in southern England, and generally attributed to *Megalosaurus*, suggest that this bulky creature walked, probably slowly and rather clumsily, on two legs. Its powerful hinged jaws and its curved, serrated teeth indicate that it was a strong predator that probably fed on a wide variety of animals, including large sauropods.

Megalosaurus's evolutionary relationships are still something of a mystery to paleontologists. The few analyses that have been done place *Megalosaurus* outside the groups that include the allosaurids and the coelurosaurs and closer to an assemblage that includes the spinosaurids (including *Spinosaurus* and *Baryonyx*). However, even this possible close relationship with spinosaurids has not yet been definitively established.

Ornitholestes

ORN-ith-oh-LESS-tees

I n 1900 a group of paleontologists from the American Museum of Natural History discovered the first—and so far the only—remains of the small theropod *Ornitholestes* in the famous Jurassic dinosaur beds at Bone Cabin Quarry in Wyoming in the western United States. Three years later, Henry Fairfield Osborn, who would later become the

skull showing possible horn

director of the American Museum of Natural History, described and named the dinosaur. He called it *Ornitholestes* because he speculated that it would probably have been a good bird catcher. This idea was further reinforced some time later when the famed painter of dinosaurs Charles Knight depicted *Ornitholestes* grabbing the earliest known bird, *Archaeopteryx* (which was contemporary with *Ornitholestes* but from a different part of the world), with its hands. There is no evidence to suggest that *Ornitholestes* did catch birds; at the same time, there is no reason to suppose that it did not. The

FIELD NOTES

■ Bird robber

6 feet 6 inches (2 m)

■ Late Jurassic

■ North America

🏛 American Museum of Natural History, New York, USA

uncertainty is compounded by the fact that Osborn seems to have mistakenly associated a hand from the same site, but from a different species, with the skull and partial skeleton of *Ornitholestes*. This hand had three fingers with sharp, curved claws, which seemed well suited to holding small prey.

The single existing specimen of *Ornitholestes*—consisting of an almost complete skeleton and a complete, but compressed, skull—is on display in New York at the American Museum of Natural History. Some reconstructions of this dinosaur show *Ornitholestes* with a thin horn at the tip of its snout, right above the nose. No horn is evident on the existing specimen, but as the tip of the snout is damaged, it is quite possible that there was a horn there.

Ornitholestes eating a salamander.

Compsognathus

KOMP-sog-NAY-thus

Compsognathus, one of the smallest known nonavian dinosaurs, was about the size of a present-day turkey. Its name refers to the delicacy of its jaw, but its fossils suggest that it was a delicate animal overall. There are two known skeletons of Compsognathus. The first, found in 1859, was from the late Jurassic lithographic limestone of southern Germany. This animal lived alongside Archaeopteryx. It lay on a slab of stone, its legs nearly perfectly preserved and its last meal, a lizard, in its ribcage. The second skeleton was found near Canjuers, in France.

Compsognathus may have had two-fingered hands. While the French specimen had three metacarpal bones (bones in the palm), it is not clear that all three had phalanges (finger bones) attached. Some paleontologists argue that this small, agile predator was—like tyrannosaurids, but unlike any other theropods—effectively two-fingered. However, because the hands

arm and leg skeletal structure

FIELD NOTES

■ Delicate jaw

🦖 6 feet (2 m)

■ Late Jurassic

■ South-western Germany; France

🏛 Bavarian State Museum of Paleontology, Munich, Germany

were not intact, it is possible that the phalanges from the third fingers were either not found or not identified. The hand of the French specimen is too poorly preserved to provide any conclusive evidence. What is certain is that Compsognathus's closest relative, Sinosauropteryx, possessed three-fingered hands.

Compsognathus has long played a central role in studies of bird origins. Because it was found in the same deposits as Archaeopteryx, and was roughly the same size, it provides an easy comparison between a primitive coelurosaur and a primitive bird. During the 19th century, similarities between these two fossils were often cited as evidence of the dinosaur–bird link. Like Sinosauropteryx, Compsognathus may well have had short, fibrous, featherlike structures on its body, although there is no direct evidence of these in the known specimens.

Michael Crichton's novels Jurassic Park and The Lost World gave Compsognathus a venomous bite that allowed groups of these fictional dinosaurs to overwhelm larger prey. It is much more likely, though, that this small animal confined itself to catching small victims and that it used its clawed fingers and toothy jaws to do so.

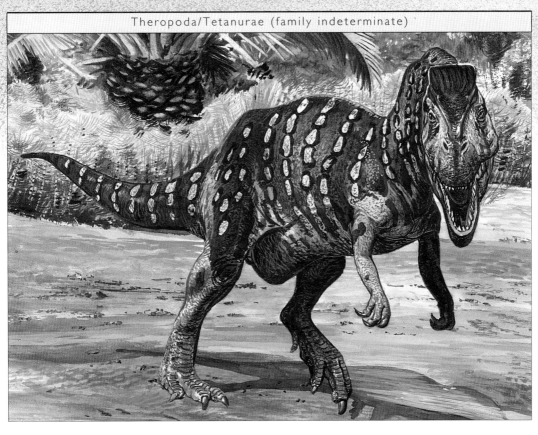

Cryolophosaurus

KRY-oh-loh-foh-SAW-rus

The geologist David Elliott discovered the remains of *Cryolophosaurus*, a moderately large theropod, in 1990. They were high up near the summit of Mount Kirkpatrick in the central Transantarctic Mountains—what would have been, in the early Jurassic, the eastern side of the great southern continent of Gondwana. The remains, which were mixed with those of a prosauropod dinosaur, a pterosaur, and a tritylodont mammal-like reptile, were excavated the following year by Dr. William Hammer. The teeth of two other kinds of theropods that were found in the sediment enclosing the remains suggest that the dead *Cryolophosaurus* had been gnawed by scavengers.

Cryolophosaurus is known from much of the skull and from a partial skeleton. It is particularly significant because it is the only relatively well-preserved theropod that has been found from the eastern part of Gondwana.

FIELD NOTES

- Frozen-crested lizard
- 24 feet (7.5 m)
- Early Jurassic
- Central eastern Antarctica
- Not on display

The features that most clearly distinguished *Cryolophosaurus* were the unusual backward-sweeping crests of bone that protruded on the top of the skull above the eyes. Small horns were situated adjacent to these crests. The crests were formed by the lacrimal bone of the skull, which extended along the entire width of the head from between the eyes. As with other crested theropods, such as *Dilophosaurus*, the crests may have been used during courtship displays or as a means of signaling to other members of the species. Despite displaying a number of primitive theropod features, *Cryolophosaurus* was closely related to the allosauroids of the later Jurassic period. Its serrated, dagger-like teeth were clearly those of a carnivore.

A side-on view of the incomplete, but reasonably well-preserved, skull of Cryolophosaurus that was discovered in 1990 in Antarctica.

Archaeopteryx

AH-kee-OP-tuh-rix

Possibly the most important fossil ever found, *Archaeopteryx* combines two rare features. First, five of the seven known specimens are preserved with impressions of the delicate feathers that it had in life. This is remarkable because generally feathers would not be strong enough to withstand the rigors of fossilization. Second, because *Archaeopteryx* displays an unambiguous mix of characters from two linked groups of animals—the birds and the dinosaurs—it is a classic and rare example of an organism on an evolutionary pathway between the two.

Archaeopteryx was a small birdlike dinosaur, about the size of a present-day crow. Its skeleton was very similar to that of some theropod dinosaurs—it had a four-toed foot with the first toe reversed to the other three; a three-fingered hand;

FIELD NOTES

■ Ancient feather

🦖 2 feet (60 cm)

■ Late Jurassic

■ Bavaria, Germany

🏛 Humboldt Museum, Berlin, Germany; Natural History Museum, London, UK

The wing of a bat (top) is made of skin and supported by all five fingers. The wings of Archaeopteryx (center) are made of feathers and arranged similarly to those of a modern pigeon (right).

a long, straight, bony tail; teeth; curved claws on the hands and feet; and a large crest on the upper arm bone.

The most strikingly birdlike feature of *Archaeopteryx* is the feathers. Not only did this dinosaur clearly have feathers, but these were arranged in exactly the same pattern as feathers on the wings of modern birds. Furthermore, they were shaped just like the flight feathers of birds that are capable of powered flight. From this we can probably assume that *Archaeopteryx* was capable of powered flight, but it may have been restricted to short bursts from tree to tree. The areas for attaching the muscles that modern birds need for flight are not very well developed in *Archaeopteryx*. It may well have used both flapping and gliding to move through the air. It probably fed on insects that it found in trees or caught in flight.

Archaeopteryx played a pivotal role in the acceptance of evolution as a mainstream

scientific theory. When Charles Darwin published *On the Origin of Species* in 1859, a perceived weakness in his argument was the lack of intermediate animals in the fossil record. If animals and plants have been changing from one form to another through time, as evolution suggests, then at least there should be some fossils of organisms intermediate in structure between different groups.

The first skeleton of *Archaeopteryx* was found just two years after the publication of Darwin's theory and, as predicted, it displayed a mix of bird and dinosaur features. Clearly, evolutionists argued, it was an intermediate form between the two groups.

Since its initial discovery there has been some debate about what kinds of reptiles *Archaeopteryx* is most closely related to. We now recognize that its closest relatives are some theropod dinosaurs, such as the dromaeosaurs and the oviraptors. In fact, the skeleton of *Archaeopteryx* is so theropod-like that one specimen found without feathers was for many years mistakenly identified as the small theropod *Compsognathus*.

As well as the seven complete or partial *Archaeopteryx* skeletons, some isolated feathers have also been found.

The tail of the Humboldt Museum specimen (above) clearly reveals the very fine lines of the barbs coming off the shaft of the feathers. All seven of the known Archaeopteryx specimens were found preserved in lithographic limestone quarried in Bavaria, Germany (left).

The superb state of preservation of the delicate, hollow bones and fine feathers was possible only because the fossils occurred in a very fine-grained limestone called lithographic limestone. This was quarried in Bavaria in order to make lithographic printing plates. It was during these quarrying procedures that the specimens of *Archaeopteryx*, as well as thousands of other important fossils, were recovered.

131

Anchisaurus

an-kee-SAW-rus

*A*nchisaurus was the first dinosaur to be discovered in America. In 1818, a small skeleton of *Anchisaurus* was found in Connecticut sandstone that was being quarried as a building material. At first it was thought to be the remains of a human and it was not until 1885 that it was recognized as a dinosaur. It is one of the many dinosaurs first described by the American paleontologist Othniel Charles Marsh.

Anchisaurus had a particularly small head and, by prosauropod standards, was very small—it was hardly bigger than a present-day sheep. Its body, however, was longer, in terms of its overall size, than those of other prosauropods. Its relatively long neck and tail, strong back leg and weaker front legs, and the defensive banana-shaped claw on its thumb were, however, typical prosauropod features. And, like most of its relatives, *Anchisaurus* could rear up on its back legs when necessary, although it would have spent most of its time walking around on all four legs. It had rounded teeth that were well

FIELD NOTES

■ *Near lizard*

🦖 *8 feet (2.4 m)*

■ *Early Jurassic*

■ *Connecticut, Massachusetts, USA*

🏛 *South African Museum, Cape Town, South Africa*

suited to grinding up the plants that it ate.

Anchisaurus lived in New England at a time when the Atlantic Ocean was just beginning to form. This area was a rift valley, similar to the Great Rift Valley in Africa today. The warm, wet climate was favorable to the growth of huge fern forests. These forests would have afforded *Anchisaurus* protection from predators and provided the dinosaur with an abundant supply of food.

Other small prosauropods that were closely related to *Anchisaurus* have been found in Europe and Africa, while fragments from Australia and Asia may also belong to very close relatives.

Anchisaurus footprints in Connecticut sandstone

Massospondylus

MASS-oh-SPON-die-lus

A medium-sized prosauropod, *Massospondylus* had a particularly small head. Its upper jaw jutted out beyond the bottom jaw and its mouth contained a variety of tooth types. According to some researchers, these features indicate that *Massospondylus* was a carnivore. They claim that it used its sharper front teeth to rip flesh from its victims and its flatter rear teeth to chew up this flesh. Most paleontologists, however, argue that *Massospondylus* was a plant-eater—like all its close relatives. The fact that grinding stones have been found from the stomach of *Massospondylus* lends support to this view. It has also been suggested that the lower jaw may have had a horny beak to bring it into line with the protruding upper jaw, but there is no physical evidence for this in the fossils.

Massospondylus was lightly built for a prosauropod, and much of its length was accounted for by its long neck and tail. Its body was similar in size to that of a large dog. The enlarged sickle-like thumb claws seen in other prosauropods were particularly well developed. We do not know exactly how *Massospondylus* used these huge claws. They would no doubt have made formidable weapons when the animal reared up on its hind legs to ward off predators—or perhaps to fight rival member of the species for mates. They may also have been useful in gathering food.

The English anatomist and paleontologist Sir Richard Owen first described *Massospondylus* in 1854. He based his description—and named the dinosaur—on the evidence of several large vertebrae. Although the earliest specimens were from southern Africa, in present-day South Africa, Lesotho, and Zimbabwe, it was later found in Arizona in the United States. Such a transatlantic distribution was possible in the early Jurassic, when the continents were joined and dinosaurs could cross what are now wide oceans.

FIELD NOTES

- Massive vertebra
- 16 feet 6 inches (5 m)
- Early Jurassic
- Arizona, USA; South Africa; Zimbabwe; Lesotho
- South African Museum, Cape Town, South Africa; National Museum of Zimbabwe, Harare, Zimbabwe

the enlarged thumb claw

133

Brachiosaurus

BRAK-ee-oh-SAW-rus

FIELD NOTES

■ Arm lizard

🦕 82 feet (25 m)

■ Late Jurassic to early
Cretaceous

■ Colorado, Wyoming, Utah, USA

🏛 The Field Museum of
Natural History, Chicago,
USA

One of the tallest known dinosaurs to walk on all fours, *Brachiosaurus* towered at least 52 feet (16 m) into the forest canopy, where its small head could strip leaves from tree branches. Some finds in the western United States suggest that *Brachiosaurus* may have been able to reach much higher. If that is true, it would have been one of the largest of all dinosaurs.

Unusual among the long-necked sauropod dinosaurs, *Brachiosaurus* had front legs that were much longer than its hind legs and a relatively short tail. The long forearms pushed the shoulders high above the level of the hips, producing the characteristic slope of its back. All of its four legs were straight, columnlike pillars that could support the stupendous weight of the animal.

Compared to the rest of the animal, *Brachiosaurus*'s head was rather small, and its mouth must have been kept busy collecting enough food to keep the creature alive. Some estimates suggest that, if *Brachiosaurus* were warm-blooded, it would have needed 440 lb (200 kg) of food a day; if, however it were cold-blooded, it would have needed much less. Its digestion was aided by grinding stones, or gastroliths, that helped to stir the soupy mixture of food that was constantly brewing in its vatlike foregut.

Its sheer size would have protected an adult *Brachiosaurus* from predators, but juveniles would have been vulnerable to attack. It is likely that this dinosaur formed small herds, where the larger individuals could protect smaller animals from the menace of predators.

The specimens of a complete skeleton once referred to as *Brachiosaurus* recovered by German paleontologists from Tenduguru, Africa, are now renamed as *Giraffatitan*. One of these specimens is a prized display in the Humboldt Museum in Berlin. An American specimen, collected from Colorado, is the largest mounted dinosaur skeleton in North America. It is on display in the entrance hall of the Field of Natural History Museum in Chicago.

skull

134

Camarasaurus

KAM-uh-ruh-SAW-rus

When the noted American paleontologist Edward Drinker Cope described *Camarasaurus* in 1877, he was obviously impressed by the hollow, box-like vertebrae in the neck. This feature made the neck much lighter and easier for the animal to carry, and it is this characteristic that gave the animal its name: "Chambered lizard."

Camarasaurus was a stout, compact sauropod with a relatively short neck and short tail. The front legs were slightly shorter than the back legs. The head can be described as a bubble of air encased by thin struts of bone. Huge holes for the nostrils, eye sockets, and other skull cavities made the skull as light as possible but strong enough to withstand the bite forces from the doglike snout.

The teeth were stumpy but strong—much more robust than the peglike teeth of other sauropods—and the jaws had a powerful bite. *Camarasaurus* was probably capable of dealing with a wider variety of tough plants, thereby giving it an advantage in mixed forests.

Camarasaurus is the most common sauropod in North America. A number of complete skeletons have been recovered, as well as numerous partial skeletons and isolated bones. Several specimens can be seen in the rocks of the Dinosaur National Monument in Utah.

As a result of these fossil finds, we now know more about *Camarasaurus* than we know about any other of the sauropod dinosaurs.

FIELD NOTES

■ Chambered lizard

🦕 59 feet (18 m)

■ Late Jurassic

■ Colorado, Utah, Wyoming, USA; Portugal

🏛 Natural History Museum, London, UK; Peabody Museum of Natural History, New Haven, Connecticut, USA; National Museum of Natural History, Smithsonian Institution, Washington DC, USA; Carnegie Museum of Natural History, Pittsburgh, Pennsylvania, USA

skull

A Camarasaurus fossil in the Devil's Canyon Science and Learning Center, Fruita, Colorado.

135

Diplodocus

dip-LOH-doh-kus

It is mainly thanks to the efforts of one man, the American industrialist Andrew Carnegie, that *Diplodocus* is now so well known throughout the world. Carnegie, a wealthy man, had a particular interest in dinosaurs, and in the late 1800s and early 1900s, he contributed liberally to the cause of paleontology. He financed many expeditions and excavations and filled his own museum—the Carnegie Museum of Natural History in Pittsburgh—with fossil skeletons from all over the United States. When his paleontologists discovered a complete skeleton of *Diplodocus*, he was so impressed with its size that he commissioned the making of 11 copies of the entire skeleton, which he gave to major museums around the world.

For many years *Diplodocus* was the longest of all known dinosaurs—it is still longer than any other dinosaur that we know from a complete skeleton. Most of its immense length is accounted for by its very long neck and

FIELD NOTES

■ Double beam

🦕 90 feet (27 m)

■ Late Jurassic

■ Wyoming, Colorado, Utah, USA

🏛 Natural History Museum, London, UK; National Museum of Natural History, Paris, France; Carnegie Museum of Natural History, Pittsburgh, Pennsylvania, USA

One fossil trackway seems to show only the front footprints of Diplodocus. It must have been floating in water and pushing itself along with its front feet.

particularly long tail. It had a small, horselike head, and its peglike teeth were restricted to the front of its mouth. *Diplodocus* used these teeth to strip large quantities of leaves from trees. It then swallowed them whole to await further processing in its huge gut. The last third of its tail was very thin and whiplike. The vertebrae at the end of the tail were reduced to simple rods.

Diplodocus's front legs were shorter than its back ones. This meant that its hips were higher than its shoulders and that its back sloped forward. It is possible that *Diplodocus* could rear up on its hind legs so that the head could reach high into the forest canopy in search of leaves. However, in the opinion of some paleontologists, this dinosaur, given its size and body structure, could not have held this pose for very long. It may well have reared up, but only in order to push over trees so that it could feed on their leaves closer to ground level.

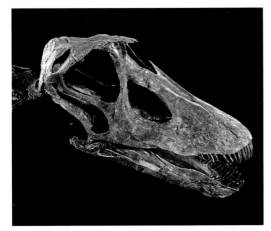

This Diplodocus skull, recovered from the Morrison Formation in Wyoming, USA, clearly shows the sharp front teeth that this dinosaur used to grasp and shred its diet of leaves and plants.

Diplodocus's curious name—meaning "double beam"—is derived from the bones on the underside of its tail, known as chevrons. In most other dinosaurs, these are simple V-shaped elements, but in *Diplodocus* they are like side-on Ts, projecting both to the front and back.

Scientists used to think that, like other sauropods, *Diplodocus* was a lumbering beast that dragged its tail along the ground. The 1980s, however, brought a renaissance in our understanding of dinosaurs. It dawned on people that, although there was ample fossil evidence of sauropods walking across ancient landscapes, there were never any impressions of tails on the ground. The only conclusion was that the tail must have been held high. However, how an animal of *Diplodocus*'s weight and dimensions managed this remained a mystery. To hold aloft a tail that measured almost half its entire body length must have cost *Diplodocus* a huge effort. The answer was in the tail's structure. Examination of *Diplodocus* skeletons revealed that massive tendons ran from the back of the head right to the tip of the tail. This tendon balanced the tail against the weight of the neck and enabled it to be held out straight behind.

This rethinking about *Diplodocus*'s posture was followed by a revision in museum displays. Around the world, copies of the *Diplodocus* skeleton that Andrew Carnegie had given out decades earlier were taken apart and reconstructed in what was now considered to be their authentic pose.

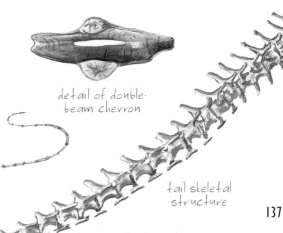

detail of double-beam chevron

tail skeletal structure

137

Barosaurus

BAH-roh-SAW-rus

Visitors to the American Museum of Natural History are greeted by a startling sight. Towering above them is a skeleton of a female *Barosaurus* protecting its infant from the menacing approach of an *Allosaurus*. The reconstruction is, of course, based to some extent on guesswork, but it is a striking depiction of how life might have been 150 million years ago.

Barosaurus is very similar to *Diplodocus*, to which it is closely related. Both were very long animals with relatively compact bodies that supported long necks and tails. Like

FIELD NOTES

- Heavy lizard
- 80 feet (26 m)
- Late Jurassic
- South Dakota, Utah, USA; Tanzania
- American Museum of Natural History, New York, USA; Utah Museum of Natural History, Salt Lake City, USA; Dinosaur National Monument, Jensen, Utah, USA

Diplodocus, Barosaurus had front legs shorter than the hind ones. As a result, its back sloped gently forward. The neck projected around 30 feet (9 m) in front of the shoulders.

It was once thought that sauropods' long necks allowed them to feed high in tree tops. However, as the neck vertebrae would not have allowed much up-and-down movement, but would have permitted a considerable sweep from side to side, we now think that these dinosaurs fed much closer to the ground on ferns and cycads.

Despite more than a century of searching and the recovery of five partial skeletons, some of them almost complete, the head and the tip of the tail of *Barosaurus* have never been found. The only clues we have about the head are a few bones from the skull collected in Tanzania and comparisons with close relatives such as *Diplodocus*. These indicate that *Barosaurus* had a horselike skull with a long snout and teeth restricted to the very front of the mouth.

The dramatic mother and infant reconstruction of Barosaurus in the American Museum of Natural History in New York.

Mamenchisaurus

mah-MEN-kee-SAW-rus

The longest neck of any animal known to us from any time belonged to *Mamenchisaurus*. It made up half the animal's total length. Reaching perhaps 49 feet (15 m) long, this incredible structure was supported by 19 vertebrae—no other dinosaur had as many neck vertebrae. Because these vertebrae were hollow—and in places the bone was as thin as egg shells—the neck was very light. Long bony struts running between the neck vertebrae would have limited its flexibility, and many reconstructions of *Mamenchisaurus* show it with the neck held straight as a ramrod. Some of these bony struts would have overlapped three or four vertebrae.

Only a few skull fragments have been found of *Mamenchisaurus*. These suggest that it had a relatively short snout with robust, blunt teeth in the front and along the sides of the mouth. The particularly heavy teeth give a clue to its diet. These teeth could have dealt with the

FIELD NOTES

■ Mamen Brook lizard

🦖 92 feet (28 m)

■ Late Jurassic

■ Szechuan, Gansu, Xinjiang, China

🏛 Beijing Natural History Museum, Beijing, China;
Museum Victoria,
Melbourne, Australia

coarser, harder parts of plants and would have been especially good for shredding cycads and other fibrous fronds.

Although superficially *Mamenchisaurus* looks similar to North American sauropods such as *Diplodocus* and *Apatosaurus*, we now think that it was part of a group of sauropods unique to Asia. By the late Jurassic, the early Atlantic Ocean had become wide enough to restrict the flow of animals between North America and Europe and, although Europe and Asia were connected by land, there were deserts and mountain ranges that would have restricted the movement of large land animals from east to west.

Mamenchisaurus was described in 1954 by the Chinese paleontologist Chung Chien Young. Young is regarded as the founder of Chinese vertebrate paleontology, and he named many dinosaurs, including *Lufengosaurus* and *Omeisaurus*.

A long neck allowed
Mamenchisaurus *to feed from tree tops.*

Apatosaurus

uh-PAT-oh-SAW-rus

*A*patosaurus is not as well known as *Brontosaurus*, but they are one and the same animal. *Brontosaurus* was well known because several relatively complete skeletons of gigantic proportions had been found and displayed worldwide. However it was later found that the fossils named *Brontosaurus* were identical to those of *Apatosaurus*, which had been named earlier. Under the rules of nomenclature, an animal can have only one name—the first published one. *Brontosaurus*, therefore, is no more.

Apatosaurus was as long as a tennis court, though still shorter, but more heavily built, than its close relative *Diplodocus*. Proportionally, its neck was shorter than that of

FIELD NOTES

■ Deceptive lizard

🦕 69 feet (21 m)

■ Late Jurassic

■ Colorado, Utah, Wyoming, Oklahoma, USA; Baja California, Mexico

🏛 American Museum of Natural History, New York, USA; Carnegie Museum of Natural History, Pittsburgh, Pennsylvania, USA; Dinosaur National Monument, Jensen, Utah, USA

Former reconstructions were wrongly given the skull of Camarasaurus *(top). The correct skull is shown at left.*

Diplodocus, and the body was not as compact, but it still had similar robust, columnlike legs. Its front legs were shorter than its back ones. As was the case with *Diplodocus*, its head had a long snout, and all the teeth were at the front of the mouth. As with other sauropods, we can only puzzle at how such a small head and tiny mouth could have gathered enough food to feed an immense body. Compounding this problem, the teeth could not chew the food. They merely crushed it to extract as much energy as possible. It seems that *Apatosaurus* and its kin spent long periods stripping leaves and other food from plants, which they swallowed whole. These vast quantities of food were held in a vatlike foregut where they stewed up, breaking down to release their nutrients. At least some sauropods swallowed stones that were held in the foregut to help stir and grind the food further.

Despite its huge size, *Apatosaurus* was controlled by a tiny brain no bigger than a cat's. This brain made up just 0.001 percent of the animal's 25-ton (25.5 t) mass. This compares with 2.5 percent in humans.

Scutellosaurus

skoo-TELL-oh-SAW-rus

Scutellosaurus is one of the earliest representatives of the armored dinosaurs (the thyreophorans) that would later include such giants as *Stegosaurus* and *Ankylosaurus*. However, compared to some of its later relatives, *Scutellosaurus* was small and lightly armored. Along its back and extending onto the base of the tail were rows of small bony "shields" embedded in the skin. Some of these shields were flat, while others were pitched, like little roofs. The largest shields formed two rows that ran along the middle of the back.

skin detail

Spread across the back, the shields formed a protective armored layer that would have defended the animal from attacks by such meat-eating dinosaurs as *Dilophosaurus*, with which it shared its early Jurassic world. *Scutellosaurus* would also have been reasonably fleet of foot and therefore able to escape predators by weaving its way through tangled undergrowth.

FIELD NOTES

- Small-shield lizard
- 4 feet 4 inches (1.3 m)
- Early Jurassic
- Arizona, USA
- 🏛 Museum of Northern Arizona, Flagstaff, Arizona, USA

Scutellosaurus fossils are found in the Kayenta Formation of Arizona. These rocks were deposited by periodic floods that spilled over river banks, but there are also the remains of sand dunes in the area, indicating that *Scutellosaurus* probably lived in an arid or semi-arid environment.

The ancestors of *Scutellosaurus* were bipedal. This feature is reflected in *Scutellosaurus* in its well-developed hind legs. However, the increased weight of the long body resulting from the armored shield would have forced *Scutellosaurus* onto all fours for most of the time. It seems likely that it could still have rocked back onto its hind legs to make a speedy getaway or to reach up and gather the higher plants that it fed on. The long tail, which made up about half its total body length, would have helped it to maintain its balance during such maneuvers.

The teeth of *Scutellosaurus* were leaf-shaped and had serrated edges. These would have been used to snip off leaves, but the lack of wear on the teeth indicates that this dinosaur did not chew its food before swallowing it.

141

Stegosaurus

STEG-oh-SAW-rus

One of the most famous dinosaurs of all, *Stegosaurus* is striking, first of all for its great size. Even more remarkable, though, are the bizarre plates and spikes that stand up, almost vertically like battlements, along each side of its backbone, from the neck right down to the middle of the tail. These plates have fascinated both paleontologists and the general public since the first specimen was described by Othniel Charles Marsh in 1877.

Marsh originally thought that these plates lay flat on the back, forming a kind of roof—hence the animal's name. Later finds showed that the plates stood upright, and that the name, which nevertheless has endured, is a misnomer. There has been ongoing debate about whether the two rows were paired, mirror images of each other, or whether they were staggered and alternating. More recent finds confirm that the plates were staggered.

But what function did these plates serve? They could have been for protection, making it difficult for an attacker to bite into its back. Channels in the plates that held blood vessels suggest that they

FIELD NOTES

■ Roofed lizard

■ 29 feet 6 inches (9 m)

■ *Late Jurassic*

■ *Colorado, Utah, Wyoming, USA*

🏛 American Museum of Natural History, New York, USA; Dinosaur National Monument, Jensen, Utah, USA; Senckenberg Nature Museum, Frankfurt, Germany; Denver Museum, Colorado, USA

may have helped to regulate the animal's body temperature. By turning its plates broadside to the sun, *Stegosaurus* could have warmed its blood; by standing with them edge-on to the sun, it could have cooled the blood down. Yet another possibility is that the plates were a display feature. Perhaps, when flushed with blood, they could change color, to impress a potential mate or to scare off a predator. It is possible that the plates in fact performed all of the above functions.

The spikes at the end of the tail—which could be more than 3 feet (1 m) long—would

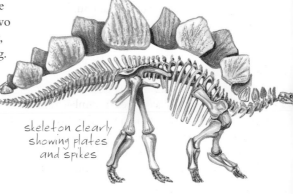

skeleton clearly showing plates and spikes

have been a very effective weapon. Swinging on the end of the long, flexible tail, they could strike a lethal blow to a potential predator. Most species of *Stegosaurus* had four tail spikes, but one species may have had eight. Unlike the staggered plates, the spikes were arranged in matched pairs.

tip of tail spikes

dorsal plates

lower back plates

The bulk of *Stegosaurus*'s weight was carried by the heavily built hind legs. These were almost twice as long as the front legs. The shorter front legs meant that *Stegosaurus* walked with its shoulders, neck, and head close to the ground.

Stegosaurus had a small head with a long snout. Its teeth were small, but wear that is evident on fossil remains indicates that the animal ground the upper and lower teeth against each other to cut and slice food. All *Stegosaurus*'s teeth were at the rear of its mouth. At the front it had a horny beak that could cut through plants as effectively as a pair of shears. Recent research suggests that *Stegosaurus* had cheek pouches in which it could hold food that was waiting to be chewed.

Not surprisingly, given the interest it has aroused, *Stegosaurus* has generated a number of dinosaur myths. One of them was the "two brain" theory. A prominent cavity in *Stegosaurus*'s hips suggested to some researchers that this "pocket" housed an auxiliary brain that controlled the rear end of the animal. We now understand that this area housed a bundle of nerves (a ganglion) that acted as a relay center that passed on messages from the brain. It has also been suggested that the animal could move its plates—that it could "wag" them as part of a display to mates or predators. But there are no scars on the skeletons that would mark the places where the huge muscles required to move the plates could have been anchored. Some paleontologists have suggested that *Stegosaurus* could stand up on its hind legs, either to reach higher food or to intimidate rivals and attackers. This now seems unlikely.

As with most dinosaurs, ideas about *Stegosaurus* have been revised in recent decades. We once thought, for example, that it dragged its tail along the ground. Now we think that it held this tail high.

ferns that *Stegosaurus* may have eaten

Kentrosaurus

KENT-roh-SAW-rus

Among the many important and beautifully preserved dinosaurs to be recovered from the Tendaguru Hills dinosaur find in Tanzania were hundreds of bones of the armored dinosaur *Kentrosaurus*. Although it is superficially similar to its larger American cousin *Stegosaurus*, there are many subtle differences between the two. *Kentrosaurus* is probably more closely related to the contemporary Chinese stegosaur *Tuojiangosaurus* and the slightly younger *Wuerhosaurus*.

Kentrosaurus had as many as seven pairs of spikes extending from the end of the tail, up over the hips, and onto the lower back. In front of the spikes were seven pairs of plates extending forward to the neck. The plates were smaller and much narrower than those of *Stegosaurus*. Both the plates and spikes were arranged in two rows and appear to have been paired rather than staggered. There was also a pair of spikes on the shoulders.

FIELD NOTES

■ Spiky lizard

🦕 16 feet (4.9 m)

■ Late Jurassic

■ Tanzania

🏛 Natural History Museum, Humboldt University, Berlin, Germany; Institute and Museum for Geology and Paleontology, University of Tübingen, Germany

Kentrosaurus was named by the German paleontologist Edwin Henning in 1915, as World War I raged. It is ironic that it became one of the many fossil victims of World War II. One of the two skeletons of *Kentrosaurus* that were pieced together from the hundreds of bones taken back to Germany was on display in the Humboldt Museum in Berlin and was destroyed during Allied bombing of the city. It has since been replaced by a copy of the second skeleton, which is still on display in Tubingen. The illustration is based on this skeleton but recent evidence from China places the pelvic spike now on the shoulder.

Kentrosaurus fed close to the ground, using a cropping beak at the front of the mouth and small chomping teeth farther back. It lived on the banks of the estuary of a huge river system.

Brain casts taken from both the Kentrosaurus fossils that have been discovered.

Tuojiangosaurus

toh-HWANG-oh-SAW-rus

Could *Tuojiangosaurus* be a dragon? Fossils of two partial skeletons of *Tuojiangosaurus* have been recovered from Szechuan in China, an area rich in dinosaur fossils. Records dating back to AD 265 tell of local villagers collecting these fossils and selling them as dragon bones for use in traditional medicines. With such a long history of fossil collecting in the area, it seems likely that fossils of *Tuojiangosaurus* were picked up and sold as parts of a mythical ancient dragon.

Tuojiangosaurus is a stegosaur, a member of the group of armored dinosaurs that includes *Stegosaurus, Kentrosaurus,* and *Wuerhosaurus.* Like all stegosaurs, *Tuojiangosaurus* had two rows of plates and spikes along its back and tail. The 17 pairs of plates were relatively small. The largest were pointed and located over the hips, becoming progressively smaller toward the neck, where they were more rounded in shape. The rows of plates petered out halfway along the tail, which was tipped by four stiletto-like spikes.

FIELD NOTES

■ Tuo River lizard

🦖 23 feet (7 m)

■ Late Jurassic

■ Szechuan, China

🏛 Zigong Dinosaur Museum, Szechuan, China; Shanghai Museum, Shanghai, China; Beipei Museum, Szechuan, China

detail of spiked tail

Another two stegosaurs—*Chialingosaurus* and *Chungkingosaurus*—lived in China alongside *Tuojiangosaurus* in the late Jurassic. Both of these were smaller than *Tuojiangosaurus*, and they may have coexisted by dividing up the plants they ate based on the height at which they fed. Alternatively, the smaller stegosaurs may have inhabited more enclosed forests, which would have protected them from predators, while the larger *Tuojiangosaurus* lived in a more open environment, protected by its size.

All three had different shaped plates, suggesting that they used them for sexual display in attempts to win mates of their own species.

Large parts of the skeleton of *Tuojiangosaurus* have never been found. We do not know what the front legs looked like below the elbow, and parts of the tail are yet to be discovered. But an almost complete skull tells us a lot about this Chinese dinosaur. Like other stegosaurs, *Tuojiangosaurus* had small teeth at the sides of the mouth, a relatively long, low snout, and a tiny brain.

145

Wuerhosaurus

WHERE-oh-SAW-rus

Wuerhosaurus, one of the last of the stegosaurs, lived about 20 million years after the time, in the late Jurassic, when these plant-eaters were in their prime.

It is not clear why the stegosaurs died out, but most scientists believe that their demise was related to the rise of the other main group of armored dinosaurs—the ankylosaurs. This assumes that stegosaurs and ankylosaurs shared the same niche in the environment, and that the emerging ankylosaurs developed some kind of advantage over the more primitive stegosaurs. Just what that advantage could have been, however, we can only guess at.

Wuerhosaurus was one of the most unusual members of a very unusual group. Whereas the plates on the backs of other stegosaurs were tall and triangular, those on *Wuerhosaurus* were long, low, and rectangular. And unlike the staggered arrangement of *Stegosaurus*'s plates—but like those of other Chinese stegosaurs—those on *Wuerhosaurus* were probably arranged in matching pairs.

While most researchers agree that *Wuerhosaurus*, like its close relatives, probably had four tail spikes, there is as yet no firm

FIELD NOTES

- Wuerho lizard
- 19 feet 6 inches (6 m)
- Early Cretaceous
- Xinjiang, China
- Not on display

evidence for them. We do know that the hips, like those of other stegosaurs, were fused into a single, broad, platelike structure across the rump, and that the vertebrae at the base of the tail had tall neural spines that may have helped *Wuerhosaurus* to suspend or maneuver its tail.

Lack of evidence continues to frustrate our attempts to fully flesh out this dinosaur. Up until now, only fragmentary skeletons and a few isolated bits and pieces have been found. There are, therefore, large chunks of the animal that we can only speculate about, and our total picture remains incomplete. We may soon learn more from a new species of *Wuerhosaurus* that was found in the late 1980s by joint excavations between Chinese and Canadian paleontologists.

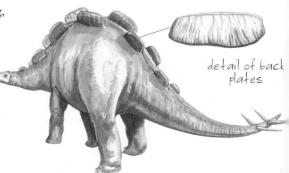

detail of back plates

Heterodontosaurus

HET-uh-roh-DONT-oh-SAW-rus

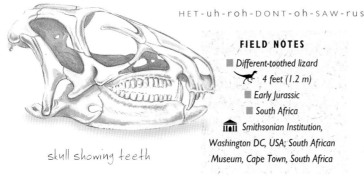

skull showing teeth

FIELD NOTES

■ Different-toothed lizard

🦖 4 feet (1.2 m)

■ Early Jurassic

■ South Africa

🏛 Smithsonian Institution, Washington DC, USA; South African Museum, Cape Town, South Africa

Living in southern Africa at the same time as *Fabrosaurus*, *Heterodontosaurus* was another of the early bird-hipped (ornithischian) dinosaurs. Like *Fabrosaurus*, it was small and fast-moving and lived on a diet of plants. In many ways, however, *Heterodontosaurus* was a more advanced animal than *Fabrosaurus*.

For one thing, it had more complex teeth. *Heterodontosaurus* gets its name from the three different types of teeth that lined its jaw. Those at the front of the mouth were small chopping teeth, while those at the back were larger, fatter, and more thickly enameled. These were suitable for grinding plant matter. Plant-eating is a challenging lifestyle—vegetable matter is often very tough and poor in nutrients. With these two types of teeth, *Heterodontosaurus* was able to eat plants more efficiently than any other kind of herbivore that had lived up to that time. It also had fleshy cheeks which helped the animal to keep extra food in its mouth while it was chewing.

Between the front and back teeth, *Heterodontosaurus* had two pairs of large tusks, similar to the canines of a carnivore. While these may have enabled *Heterodontosaurus* to eat meat as well as plants, it seems more likely that it used them for digging up roots or for displaying to rivals or to prospective mates. In some species of modern deer, only the males have tusks, but we do not know if *Heterodontosaurus*'s tusks were present in only one or in both of the sexes.

The "thumb" of *Heterodontosaurus*'s hand may have been opposable, and so it may have been used to grasp vegetation. The front legs were relatively long. *Heterodontosaurus* may have stood on all fours while grazing on ground plants, but if it needed to move quickly, it would have run on its powerful back legs.

Heterodontosaurus was one of the earliest of the ornithopod dinosaurs. Many of the features that were important to this group's later success—batteries of grinding teeth, fleshy cheeks, the ability to run on the back legs—can be seen in ancestral form in this little dinosaur.

147

Fabrosaurus

FAB-roh-SAW-rus

A contemporary of *Heterodontosaurus*, and sharing the same southern African environment in the early Jurassic, *Fabrosaurus* lived on tough plant food, such as ferns and cycads. It was one of the earliest ornithischian (bird-hipped) dinosaurs. Later ornithischians included such diverse forms as the ankylosaurs, ceratopsians, and hadrosaurs. In the much smaller *Fabrosaurus*, however, we can get a glimpse of the sort of animal from which these larger herbivorous dinosaurs evolved.

Fabrosaurus had small front limbs and very long hindlimbs and was therefore well adapted for standing and running on its back legs. Its long tail balanced the body in front of the hips, and the back legs were held vertically under the body and worked

FIELD NOTES

- Fabre's lizard
- 3 feet 3 inches (1 m)
- Early Jurassic
- Southern Africa
- Not on display

like those of a mammal, rather than like the sprawling limbs of most reptiles.

Its small size and light body structure—it had hollow bones and hollow places in the skull— meant that *Fabrosaurus* would probably have been able to flee from a predator at considerable speed.

The hands of *Fabrosaurus* had four larger fingers and a smaller fifth finger. These digits had claws rather than hooves. This dinosaur probably did not often use its front limbs for walking; they would have served mainly for holding foliage while the animal ate.

Compared to those of its later relatives, the small, pointed teeth of *Fabrosaurus* were quite simple in their structure and its jaw action was strictly up and down, allowing it to slice or cut up plant matter. *Fabrosaurus* was not as well adapted for grinding and chewing food as was *Heterodontosaurus* or the later ornithischians.

In the early 1970s, *Fabrosaurus* became widely known as *Lesothosaurus*, after the country where the first fossils were found. Both names are still in use, which is a source of some confusion. However, as *Fabrosaurus* was the first to be used, it should have precedence.

complete
skeleton

148

Dryosaurus

DRY-oh-SAW-rus

O rnithopods of the Jurassic were small to medium-sized herbivorous dino-saurs that relied on their agility and speed, rather than size, armor, or weapons to defend themselves against predators. *Dryosaurus*, was a common and widespread orni-thopod of the late Jurassic, and was fairly typical of this group.

The Yale University paleontologist and fossil-hunter Othniel Charles Marsh named *Dryosaurus* in 1894. Thanks to ex-tensive finds of adult, juvenile, and baby specimens, it is one of the best understood of all dinosaurs. It stood, and ran, on its long hind legs, its body counterbalanced by its long tail. The anatomy of *Dryosaurus*'s feet and legs shows that this dinosaur was an adept runner. According to some estimates, it could reach speeds of more than 25 miles per hour (40 km/h)—fast enough to outrun most predators.

The long shin bones in its hindlimbs helped Dryosaurus *to move at considerable speeds when it needed to escape from danger.*

FIELD NOTES

■ Oak lizard

10–13 feet (3–4 m)

■ Late Jurassic

■ Europe (England, Romania); Africa (Tanzania); North America (Colorado, Utah, Wyoming)

🏛 Carnegie Museum, Pittsburgh, USA

cycads

Dryosaurus was not only quick across the ground, it was a fast grower, too. Analyses of the bone tissue of baby, juvenile, and adult *Dryosaurus* reveal that it grew continuously. There is no sign that growth slowed down during a cold winter or a harsh dry season as it does, for example, with many present-day reptiles. Many researchers believe that only warm-blooded animals can grow continuously, so the growth pattern of *Dryosaurus* has fueled speculation that dinosaurs, or at least the small ones, were warm-blooded. Whatever the metabolic rate of *Dryosaurus*, its babies did not take long to grow to adult size.

As with many ornithopods, the fast growth rate was facilitated by an efficient battery of cheek teeth, which were thickly enameled and could grind up plant matter almost without pausing. Before grinding up this plant material, *Dryosaurus* cropped it with a sharp, horny beak—it had no teeth at the front of its mouth.

149

Tyrannosaurus

tie-RAN-oh-SAW-rus

Tyrannosaurus rex exemplifies every image conjured by the word "dinosaur": massive, ferocious-looking, and extinct. It is the only dinosaur that is commonly known by both its generic and specific names. Despite this, fossil evidence for this animal was surprisingly scant until quite recently, and it was only in the 1990s that important gaps in our understanding of Tyrannosaurus were filled in.

Tyrannosaurus was one of the largest of predatory dinosaurs. Some individuals measured as much as 42 feet (12.8 m) long and were up to 13 feet (4 m) tall at the hip, with a skull more than 5 feet (1.5 m) long. By any standards, Tyrannosaurus was a tremendous animal.

This giant was also one of the last of the nonavian dinosaurs. All the Tyrannosaurus skeletons have come from the latest Cretaceous deposits of the United States and Canada, although some researchers regard Tarbosaurus, a large tyrannosaurid from slightly older deposits of Mongolia, to have been a form of Tyrannosaurus.

FIELD NOTES

■ Tyrant lizard

🦖 40 feet (12 m)

■ Late Cretaceous

■ North America (Canada, USA)

🏛 American Museum of Natural History, New York, USA; Carnegie Museum, Pittsburgh, USA; Field Museum, Chicago, USA; University of California Museum of Paleontology, Berkeley, USA; Los Angeles County Museum of Natural History, Los Angeles, USA; Museum of the Rockies, Bozeman, USA; Denver Museum of Natural History, Denver, USA; Tyrrell Museum, Alberta, Canada; Senckenberg Museum, Frankfurt, Germany

Like other tyrannosaurids, Tyrannosaurus had two very short forelimbs and only two functional fingers on each hand. The forelimb of the longest known specimen was hardly any longer than the forearm of an adult human. The front teeth were D-shaped in cross-section, and each cheek bore 12 rather robust teeth, which were shaped more like serrated bananas than the steak-knife shapes seen in most other theropods.

Henry Fairfield Osborn first described Tyrannosaurus in 1905

Parts of a Tarbosaurus, *a close relative of* Tyrannosaurus *from Mongolia.*

Teeth of Tyrannosaurus.

from fossils that
Barnum Brown
had collected in
Montana. More fossils
that came to light in the
course of the next few years
allowed Osborn to amplify
his interpretation.

Over the years new discoveries were made,
including several more complete specimens.
However, no hand came to light until 1990,
when John Horner, of Montana State University, published an account of a *Tyrannosaurus*
specimen from Montana in which the hand
was preserved. This discovery confirmed the
presence of only two digits—something that
scientists had suspected by analogy with other
tyrannosaurids. Osborn's reconstructions showed
a three-fingered hand—a sensible guess, as all
other theropods known at the time of his
reconstructions had three fingers.

In 1991, a group of commercial fossil-hunters
discovered "Sue" on a ranch in South Dakota.
It is perhaps the largest and most complete
Tyrannosaurus skeleton ever found. A legal battle
over ownership followed this discovery. Finally
the courts awarded the fossil to the rancher, who
in 1997 sold it at auction to the Field Museum
in Chicago. Researchers have high hopes for
Sue and expect it will significantly extend our
knowledge of *Tyrannosaurus*. A high-resolution

computed tomographic (CT)
study of the skull is giving
scientists access to what were
hitherto inaccessible internal
details of this dinosaur's head.

The predatory habits of
Tyrannosaurus are still unclear.
Some people, presuming it was
a slow mover and citing the smallness
of its forelimbs, maintain that it must have
been solely a scavenger. Others, who claim
that it was an active hunter, point to the strength
of this animal's teeth and the evidence that bite
marks found on *Triceratops* bones seem to have
been made by *Tyrannosaurus* teeth.

*In May 2000, nine years after it was discovered, the
reconstructed skeleton of "Sue" finally went on display
at Chicago's Field Museum.*

Albertosaurus

al-BERt-oh-SAW-rus

Albertosaurus was a slightly older relative of *Tyranno-saurus* and was very similar in appearance. *Tyrannosaurus* lived between 70 and 65 million years ago; *Albertosaurus* roamed the North American late Cretaceous world between 75 and 70 million years ago. Like *Tyrannosaurus*, *Albertosaurus* was a huge biped with two-fingered hands and thin plates of bone covering some of its skull openings. *Albertosaurus* was smaller and had a narrower skull than *Tyrannosaurus*, and its eyes looked more to the side. In front of its eyes, *Albertosaurus* had a pair of small, blunt horns, which it may have used for sexual display. Perhaps because its body was smaller, *Albertosaurus*'s skeleton was more gracile than that of *Tyrannosaurus*. It was also a faster mover than its more cumbersome successor.

Searches of fossil-bearing beds in south-western Alberta have yielded some amazing discoveries, including a number of nearly complete skeletons of juvenile *Albertosaurus*. While adult *Albertosaurus* had rather stout bones

FIELD NOTES

■ Alberta lizard

🦖 26 feet (8 m)

■ Late Cretaceous

■ South-western Canada

🏛 Royal Tyrrell Museum, Alberta, Canada; Royal Ontario Museum, Toronto, Canada; American Museum of Natural History, New York, USA; Field Museum, Chicago, USA

Fossil foot from the American Museum of Natural History

tearing chunks of flesh with powerful jaws

in their hindlimbs, the hindlimbs of very young specimens were remarkably slender. In some cases, several *Albertosaurus* skeletons—at different stages of growth—have been found together. On the strength of these finds, some researchers have suggested that these dinosaurs traveled as family groups. However, there is no conclusive evidence that skeletons found together were from animals that actually lived together as a group.

Albertosaurus is the best known of all tyrannosaurids. Recent discoveries include wishbones—a feature it shared with other advanced theropods as well as with birds. Many museum specimens of *Tyrannosaurus* have filled in gaps in our knowledge of the larger dinosaur with information taken from *Albertosaurus*.

Giganotosaurus

JIG-an-oh-toh-SAW-rus

When the discovery of *Giganotosaurus* was announced to the world in 1995, the news caused a sensation. For almost all of the 20th century, paleontologists had believed *Tyrannosaurus* to have been the largest of all theropods. However, *Giganotosaurus*, from the late Cretaceous of southern Argentina, proved to be at least as large. In their description of *Giganotosaurus*, Argentinian paleontologists Rodolfo Coria and Leonardo Salgado reported a length of 41 feet (12.5 m)—longer than some *Tyrannosaurus* skeletons.

The first *Giganotosaurus* discovery consisted of an incomplete skeleton. Paleontologists could tell that this new dinosaur was an allosauroid—it was obviously closely related to animals like *Allosaurus* and, especially, *Carcharodontosaurus*—but because the material was so incomplete, the estimates of its size were far from conclusive. Since 1995, a number of new discoveries have been made. In one case, several *Giganotosaurus* skeletons were found

FIELD NOTES

■ Giant southern lizard

🦖 43 feet (13 m)

■ Late Cretaceous

■ Patagonia, Argentina

🏛 Academy of Natural Sciences, Philadelphia, USA

close together. This suggests that these animals may have moved around in groups.

In early 2000, a team of Argentinian and Canadian paleontologists announced the discovery of a well-preserved, slightly younger relative of *Giganotosaurus* in Argentina. The newly discovered dinosaur, named *Mapusaurus* in 2006, may have been as long as 43 feet (13 m), but was bulkier than *Giganotosaurus* and heavier than *Tyrannosaurus*.

Giganotosaurus lived alongside a number of giant sauropods. Some of these had bony plates on their backs, which may have afforded them some protection against attack from above. Only one theropod in the region was large enough to attack them from above—that predator was *Giganotosaurus*.

Size comparison of three theropods, from back to front: Giganotosaurus, Cretaceous; Torvosaurus, Jurassic; Coelophysis, Triassic.

153

Baryonyx

BARE-ee-ON-icks

FIELD NOTES

■ Heavy claw

🦕 30 feet (9 m)

■ Early Cretaceous

■ Southern England

🏛 Natural History Museum,
London, UK

The first part of *Baryonyx*—one of the best known of British dinosaurs—to be discovered was a huge claw. In 1983, the amateur fossil-hunter William Walker happened upon it in a clay pit in Surrey, in southern England. A group of paleontologists from the Natural History Museum in London later visited the site and uncovered most of the skeleton of *Baryonyx* in the early Cretaceous deposits.

Baryonyx and its relatives, the spinosaurids, were very unusual theropods. The claw that led professional paleontologists to the scene was from the hand. Although saurischian dinosaurs ancestrally had enlarged thumb claws, they were nowhere near as large as those on *Baryonyx* and other spinosaurids. *Baryonyx* had very powerful forelimbs, so much so that scientists initially assumed that it was capable of walking on all fours—an idea that has since been discounted. While all theropods had strong, grasping hands, *Baryonyx*'s hands were more massive. This suggested that it made extensive use of them for grappling with prey.

Some researchers have suggested that *Baryonyx* was principally a fish-eater and that it used its hands to grasp fish and its long, slender snout to snap prey out of the water. Living crocodiles that specialize in catching fish often have similarly slender jaws. *Baryonyx*'s teeth were less flattened than the teeth of most theropods and they were very finely serrated. This feature, coupled with the thin jaws, has led many scientists to conclude that *Baryonyx* was not capable of attacking and bringing down large animals, and so lends support to the theory of a fish diet. The fact that *Iguanodon* bones were found in *Baryonyx*'s ribcage may suggest that it set its sights on larger prey. However, there is no way of knowing whether *Baryonyx* killed this *Iguanodon*. It may simply have happened upon, and scavenged, an *Iguanodon* corpse.

This huge fossil claw of the carnivorous dinosaur Baryonyx was discovered in a Surrey clay pit in 1983.

Suchomimus

SOO-koh-MY-mus

FIELD NOTES

■ Crocodile mimic

🦖 36 feet (11 m)

■ Early Cretaceous

■ Niger, northern Africa

🏛 Children's Museum, Chicago, USA

skull showing
elongated jaw

Suchomimus was one of the largest known spinosaurid dinosaurs. It was discovered in Niger by a party led by Paul Sereno from the University of Chicago and was first described in 1998. Shortly before that, Dale Russell and Philippe Taquet announced the discovery of a very similar dinosaur from another part of Niger, but the fossils that were described by Sereno were much more complete and allowed for a more detailed reconstruction.

Like other spinosaurids, Suchomimus had a low and slender snout, much the same as that of a modern crocodile—hence this animal's name. Suchomimus and other spinosaurids had a secondary palate. The nasal passages stretched all the way to the back of the mouth cavity, as they do in living mammals and crocodiles, and did not open within the mouth, as they do in many living reptiles and most dinosaurs. This was possible because the nasal openings were set back from the tip of the snout. We do not know why these animals evolved these features, but the secondary palate may have strengthened the narrow snout or even allowed Suchomimus and its relatives to keep the tip of the snout submerged while they hunted for fishes.

Like Baryonyx, Suchomimus had pointed teeth with very fine serrations. Also like Baryonyx, it had massively built forelimbs and a large sickle-like claw on each thumb. The nature of the teeth, together with its slender, gracile snout, suggests that Suchomimus may have been unable to catch large prey and so fed largely on fishes—possibly either picking them up with its muzzle or grasping the slippery creatures with its bladelike claws.

This reconstruction of Suchomimus was for some time at the National Geographic Center in Washington DC, USA.

155

Spinosaurus

SPY-noh-SAW-rus

FIELD NOTES

■ Spiny lizard

🦖 56 feet (17 m))

■ Late Cretaceous

■ North Africa (Egypt, Niger)

🏛 Not on display

At the beginning of the 20th century, the German paleontologist Ernst Freiherr Stromer von Reichenbach, of the University of Munich, began a series of expeditions to Egypt, where he discovered, in the western Sahara, the remains of several formerly unknown late Cretaceous dinosaurs. The most significant of these finds was *Spinosaurus*. It was so named because of the tall spines on its vertebrae. Stromer collected some fragmentary remains of *Spinosaurus* in 1912 and described the dinosaur in 1914.

internals of the sail

Unfortunately, the 20th century has been unkind to this dinosaur. Stromer's specimen of *Spinosaurus*, as well as those of several other Egyptian dinosaurs he described, were destroyed in bombing raids during World War II, and much of the important information concerning specific localities was lost during both the world wars. Fortunately for science, however, Stromer published an extensive and meticulously detailed description of the material he collected.

In many ways, *Spinosaurus* resembled a number of other theropods, such as *Allosaurus*. Unlike other spinosaurids, however, it had neural spines on its vertebrae. What set *Spinosaurus* apart was the great size of these spines, which rose more than 5 feet (1.5 m) high and formed a "sail" over the back, much like that of the Permian mammal relative *Dimetrodon*. We now suspect that these spines were connected by skin, but we have no clear idea of how the sail functioned. It could have helped *Spinosaurus* to radiate or absorb heat and thus regulate its body temperature. It may also have served as a social signaling device or as a means of sexual display.

New discoveries have shown that *Spinosaurus* could have been the biggest of all meat-eating dinosaurs, reaching 56–59 feet (17–18 m) in length. Its long slender snout had serrated crocodile-like teeth, conical and straight, unlike the curved, compressed teeth found in most other meat-eating dinosaurs.

156

Gallimimus

GAL-ee-MY-mus

allimimus was one of the largest and best known of the ornithomimid, or "bird mimic," dinosaurs. It was one of the important discoveries made by joint Polish–Mongolian expeditions to Mongolia during the 1960s, and skeletons from several stages of this dinosaur's growth have been collected. The largest were animals 10 feet (3 m) long. More than two-thirds of this length, however, is accounted for by the neck and tail.

As with other ornithomimids, *Gallimimus*'s skull was relatively small and its jaws were toothless. Its limbs and neck were very long and slender, and the bones of the palm—the metacarpals—were particularly long. Even the metacarpal for the thumb was much longer than in most theropods, where the thumb is usually markedly shorter than the other digits.

We think that *Gallimimus*, and the other ornithomimids, were capable of great bursts of speed—they were certainly depicted as fast-moving animals in the film *Jurassic Park*. *Gallimimus*'s arms had considerable freedom of movement, although they do not seem to have been able to reach very high. To compensate for this, its long neck could probably have stretched

FIELD NOTES

- Chicken mimic
- 10 feet (3 m)
- Late Cretaceous
- Mongolia
- Paleontological Museum, Mongolian Academy of Sciences, Ulan Baatar, Mongolia

down, allowing *Gallimimus* to bring food items from its hands up to its mouth.

This dinosaur may have fed by standing in shallow water and sifting food through its beak like a flamingo, as well as hunting for small insects. Its hands, which were more suited to digging than grasping, may also have scooped up the eggs of other dinosaurs from the soil and then cracked them open with the broad beak at the tip of its long snout.

Mongolian ornithomimids such as *Gallimimus* are typically found along with large tyrannosaurid theropods and duckbills, but not with the horned ceratopsians. North American ornithomimids, on the other hand, are more commonly found with diverse and abundant ceratopsians. This suggests that Asian and North American members of this group lived under different conditions. However, just what these differences were we have yet to discover.

long beak and neck

Struthiomimus

STROO-thee-oh-MY-mus

Picture a plucked ostrich with a long tail stretching stiffly out behind and a pair of human-sized arms with hands with three claws attached to them, and you have a mental image of *Struthiomimus*. This long-necked, long-legged theropod is, along with the Mongolian *Gallimimus*, the most well known member of the ornithomimid, or "bird mimic," dinosaurs.

Struthiomimus was roughly the same size as a modern ostrich and could probably reach a similar top speed of about 50 miles per hour (80 km/h) on flat, open ground. With its small and very lightly built skull, tooth-less beak, and very large eyes, it was strikingly similar to today's flightless birds. The three large, clawed, forward-facing toes on each of its feet were typical of theropod feet, but they were also very birdlike. Ornithomimids lacked true feathers and may have had naked

skeleton

FIELD NOTES

- Ostrich mimic
- 13 feet (4 m)
- Late Cretaceous
- Alberta, Canada
- Royal Tyrrell Museum, Alberta, Canada

skin. However, some juvenile specimens display a partial covering of downlike filaments similar to those discovered on other birdlike dinosaurs.

The lightly built head and toothless beak imply that *Struthiomimus* could not have killed or eaten large animals, but it would certainly have been able to pursue and swallow small reptiles and large insects. With the long, hooklike claws on its hands it could either have dug small animal prey out of shallow burrows or pulled succulent leaves and primitive fruits down from low trees. By stretching its neck, it could have reached reasonably high into tree branches. Gastroliths (gizzard stones) that have been found at the front of the rib cage of *Struthiomimus* skeletons indicate that plant material was an important component of its diet. This unusual theropod was, therefore, either omnivorous or entirely herbivorous.

First described in 1902, *Struthiomimus* was long known only from a single complete skeleton and several partial skeletons. A number of complete skeleton have recently come to light.

Oviraptor

oh-vee-RAP-tuh

Oviraptor was a strange-looking animal, especially for a theropod. It had a short snout, toothless jaws, and a rounded mass of thin bone over its nose, rather resembling a chicken's comb. *Oviraptor* had slender limbs and, like its closest relatives, was probably fast on its feet. Though *Oviraptor* is known only from the late Cretaceous of Mongolia, other *Oviraptor*-like animals are known from the Cretaceous of other parts of Asia and western North America.

This dinosaur was discovered during the 1920s American expeditions to Mongolia. Most of the specimens were found near nests of dinosaur eggs. Because the eggs were thought to belong to *Protoceratops*—a small, common ceratopsian in that region—it was assumed that *Oviraptor* was stealing the eggs. The eggs, however, lacked embryos. When, in the 1990s,

FIELD NOTES

■ Egg robber

🦖 10 feet (3 m)

■ Late Cretaceous

■ Mongolia

🏛 American Museum of Natural History, New York, USA; Academy of Science, Ulan Baatar, Mongolia

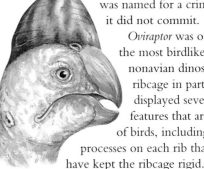

Two species of Oviraptor: *O. philoceratops (above)* and *O. mongoliensis (right)*. Dinosaur crests were very diverse and changed constantly during the animal's lifetime.

American teams returned to Mongolia, they found more of the same kinds of eggs, including some that contained the delicate skeletons of embryos—but they were embryos of *Oviraptor*, not of *Protoceratops*. Soon after that, more amazing fossils came to light. They were of *Oviraptor* skeletons sitting on nests, with their forelimbs wrapped around the eggs. These were probably skeletons of parents incubating and protecting their own eggs. *Oviraptor*, it would seem, was named for a crime that it did not commit.

Oviraptor was one of the most birdlike of the nonavian dinosaurs. Its ribcage in particular displayed several features that are typical of birds, including a set of processes on each rib that would have kept the ribcage rigid.

Recently a relative of *Oviraptor* was found with a pygostyle—a set of fused vertebrae that would later help support the tail feathers of birds.

159

Troödon

TROO-oh-don

The first fossil of *Troödon* was a single leaf-shaped tooth discovered in 1854 in the Judith River Formation in Alberta, Canada. On the basis of this tooth, Joseph Leidy described and named the animal in 1856. However, Leidy considered *Troödon* to be a lizard and not a dinosaur. Later discoveries of teeth, from the late Cretaceous of Montana, confirmed *Troödon* to be a dinosaur. For some time this dinosaur was thought to be a carnivorous ornithopod—its teeth resembled those of some ornithopods—but eventually it was confirmed to be a small birdlike theropod.

Thanks to the discovery of a number of sometimes fragmentary skulls and skeletons, we now know that *Troödon* had a long snout. However, instead of the row of long, curved teeth that were typical of most theropods, *Troödon* had a large number of relatively small teeth. Each side of the lower jaw may have had as many as 35 teeth, all of them with large

FIELD NOTES

- Wounding tooth
- 10 feet (3 m)
- Late Cretaceous
- North-western North America
- Museum of the Rockies, Bozeman, USA

serrated-edged tooth

serrations. No other theropod is known to have had so many teeth. The teeth varied in shape, depending on their location in the jaw.

Troödon is famous for the great size of its brain cavity. *Troödon* and other troödontids may have had the largest brains, relative to body size, of any nonavian theropods. This has led many scientists to conclude that *Troödon* and its close relatives were more intelligent than other dinosaurs. Such a belief is based on the observation that in living animals there is a rough correlation between relative brain size and degrees of intelligence. However, as intelligence leaves no trace in the fossil record, this must remain a matter of speculation. *Troödon*'s very large eye sockets have also led some researchers to believe that it may have been a nocturnal animal.

Recent studies suggest that the troödontids were among the closest extinct relatives of birds. They may also have been very closely related to the dromaeosaurids, such as *Deinonychus* and *Velociraptor*. The claw on *Troödon*'s second toe was longer than those on its other toes, but *Troödon* lacked the long "killer" sickle that characterized the dromaeosaurids.

Velociraptor

vel-OSS-ee-RAP-tor

Velociraptor is the night-marish villain of the movie *Jurassic Park*. Swift, birdlike and intelligent—as presented on film, it embodies all the dynamic qualities that, until quite recently, dinosaurs were thought to lack.

Although in life *Velociraptor* was most likely a swift-running predator, the real animal—from the late Cretaceous of Mongolia—differs in several significant ways from its cinematic namesake. For one thing, the skull was narrower than was depicted on screen. In fact, the heads of the movie "raptors" more closely resemble that of *Deinonychus*. The tail, too, was different.

FIELD NOTES

■ Fast robber

🦕 3 feet (1 m)

■ Late Cretaceous

■ Mongolia

🏛 Paleontological Museum, Mongolian Academy of Sciences, Ulan Baatar, Mongolia; American Museum of Natural History, New York, USA; Paleontological Institute, Moscow, Russia

Velociraptor was a dromaeosaurid, and dromaeosaurid tails were broomhandle-stiff, not the flexible structures that were seen in the movie. Most important of all is the question of their size. The *Velociraptor* of the late Cretaceous was barely longer than 3 feet (1 m); in the movie, however, it was shown to be the size of a large adult human—and it looked all the more ferocious as a result. It is interesting to note that while *Jurassic Park* was still being filmed, the remains of *Utahraptor*—a previously unknown dromaeosaurid, which was about the same size as those "faked" for the movie, were found in Wyoming.

As with other dromaeosaurids, *Velociraptor* probably used its hooked talon to kill its prey. One celebrated *Velociraptor* skeleton, found in Mongolia in 1971, seems to confirm this theory. *Velociraptor* was entangled with a *Protoceratops* skeleton, the *Velociraptor* foot claw embedded in the *Protoceratop*'s rib cage. A sand dune must have collapsed on the two animals as they were grappling with each other, smothering both and preserving this dramatic scene of predation and disaster for millions of years.

In May 1997 a replica of Velociraptor, as it appeared in the movie Jurassic Park, went on display as part of a special exhibition at the American Museum of Natural History in New York, USA.

161

Deinonychus

die-NON-ee-kus

A lthough it was not the first of the dromaeosaurs to be discovered, *Deinonychus*, which was one of the largest of them, was the first to be fully described. After it was described, the dromaeosaurids, which are now improperly known as "raptors," became recognized as some of the most chillingly efficient predators that have ever lived.

Deinonychus was an agile bipedal theropod with large, partly forward-facing eyes, a relatively large brain, and a long, narrow snout lined with recurved, serrated, bladelike teeth. *Deinonychus* was an intelligent, very well-equipped predator that used the three long-clawed fingers on each of its large hands to snatch small prey or to inflict terrible wounds on large animals. It may also have employed its claws as grappling hooks to clamber up on the bodies of larger dinosaurs that it hunted and attacked in packs.

FIELD NOTES

■ Terrible claw

🦖 10 feet (3 m)

■ Early Cretaceous

■ Montana, Wyoming, Oklahoma, USA

🏛 Peabody Museum of Natural History, New Haven, USA

Deinonychus walked, ran, and jumped mainly on the outer (third and fourth) toes of each foot. These toes had long, sharp claws. The first toe was also clawed but this claw was quite short. The claw on the second toe was, by contrast, an extra-large, curved slashing weapon up to 5 inches (13 cm) long that was able to swivel through 180 degrees. *Deinonychus* usually held this claw back off the ground in order to keep the point sharp, loaded, and "cocked" ready to go in for the kill.

Deinonychus's foot was equipped with three sharply clawed toes (right). The "terrible claw" on the second toe of each foot (above) was a highly flexible weapon.

This model of Deinonychus shows its predatory features.

As in many other dinosaurs, the tail of *Deinonychus* was stiffened for about three-quarters of its length by bundles of overlapping bony rods, which, however, were flexible at the base. This stiffening allowed the tail to be controlled by a few large muscles that connected to the hips and hindlimbs. It helped *Deinonychus* to make rapid lunges or to change direction suddenly when running— *Deinonychus* may even have been able to turn around in mid-air while leaping to catch prey. It also prevented the tail from flexing from side to side in time with the limb movements while the animal was running, thus preventing energy from "leaking away" into the tail and helping the dinosaur to run faster and more efficiently.

Grant Meyer and Professor John Ostrom of Yale University first discovered remains of *Deinonychus* in southern Montana in 1964. Excavation at this site produced remarkable finds—nearly complete skeletons of four *Deinonychus*, together with a skeleton 20 feet (6 m) long of the *Iguanodon*-like ornithopod *Tenontosaurus*. When Ostrom described and named *Deinonychus* in 1969, he suggested that *Tenontosaurus* was its preferred quarry and that this association of predators and prey was evidence for pack hunting behavior in *Deinonychus*. Certainly a single *Deinonychus* would not have been capable of killing a herbivore as large as *Tenontosaurus*, though the death of four *Deinonychyus* in the attack could hardly be considered typical.

If, as seems probable, *Deinonychus* did live in packs and hunted much larger dinosaurs, it must have been similar in behavior and ecology to present-day wolves, hunting dogs, and hyenas. This implies that *Deinonychus* was not only an acrobatic predator with an armory of deadly weapons, but also an endurance athlete—and one that existed, probably, as part of a caring, cooperative social group. Territories near the migration routes or breeding sites of large herbivores could have supported large packs of 20 or more *Deinonychus*, dominated by a few breeding individuals. Non-breeding adults would have defended the territory from rival groups of *Deinonychus* and other carnivores, and they would also have helped to feed the young of the dominant breeding animals.

Dromaeosaurus

droh-MAY-oh-SAW-rus

Although until recently *Dromaeosaurus* was known almost exclusively from a skull and some fragmentary bones found by Barnum Brown in 1914, in the Judith River Formation in Alberta, Canada, it has given its name to a family of dinosaurs—the dromaeosaurids. Similar in most ways to its better understood cousins *Deinonychus* and *Velociraptor*, *Dromaeosaurus* was a small, birdlike theropod with a claw that resembled a switchblade on the second toe of each foot. Even though new evidence of *Dromaeosaurus* has come to light in recent times, the scarceness of its fossil remains suggests that it may have been a rare theropod.

FIELD NOTES

■ Running lizard

🦖 6 feet (1.8 m)

■ Late Cretaceous

■ Alberta, Canada

🏛 Royal Tyrrell Museum, Alberta, Canada

One feature common to all dromaeosaurids was the extreme stiffness of their tails. The tails of most theropods were stiffened, at least for part of their length. This relative inflexibility was caused by the lengthening of the attachment processes (bony rods known as zygapophyses) on the vertebrae. In the dromaeosaurids, however, these rods became greatly elongated, each of them extending over several vertebrae and effectively bracing them together.

As a result of this, the thin tails of the dromaeosaurids became completely inflexible. The stiff tail would have helped *Dromaeosaurus* in pursuing prey by acting as an effective counterbalance behind the hip.

The world of *Dromaeosaurus* was filled with giant ceratopsians, duckbills, pachycephalosaurs, ankylosaurs, and tyrannosaurids such as *Albertosaurus*. An individual *Dromaeosaurus* would probably have been regarded as no more than a minor nuisance by most of these, but it would have been a source of terror to smaller dinosaurs and other vertebrates in its territory. Hunting in groups, however, may well have given *Dromaeosaurus* the capacity to target and bring down much larger prey.

The Dromaeosaurus skeleton at the Royal Tyrrell Museum in Drumheller, Alberta, Canada.

Avimimus

AY-vee-MY-mus

As its name clearly suggests *Avimimus* was a very birdlike dinosaur. It was a lightly built, long-legged animal that may have made its living by chasing after small reptiles and mammals. *Avimimus*'s diet is still a subject of debate; some scientists believe that it probably fed mainly on insects. It was capable of swift speeds, which would have both enhanced its efficiency as a predator and helped it to avoid falling prey to larger meat-eating dinosaurs.

Avimimus was a strange-looking dinosaur. It had a long neck and a short head with a tooth-less mouth. It had a large braincase, which suggests that its brain, too, was relatively large. In place of teeth it had a powerful beak, similar to that of a present-day cockatoo. The bones of its wrist were fused together as they are in birds and the three long fingers could be held tucked under the body. In fact, *Avimimus* could fold its whole arm against its body in a similar fashion to the way a bird folds its wing. Its long legs ended in three toes, with a smaller fourth toe nestled on the inside of the foot. All the toes and fingers were tipped with sharp, curved claws. Unlike

FIELD NOTES

■ Bird mimic

🦖 5 feet (1.5 m)

■ Late Cretaceous

■ Mongolia

🏛 Not on display

a bird, however, *Avimimus* had a long, bony tail and its pelvis resembled those that are seen in other theropods.

It is possible that *Avimimus* had feathers of some kind, but the deposits in which it has been discovered are too coarse for such features to have been preserved. However, a rough ridge on the forearm of *Avimimus*—similar to the one on *Caudipteryx*, which supported a half-wing—may well have served to anchor feathers. However, even if *Avimimus* did have feathers, it seems unlikely that it would have been capable of flight.

We have known about *Avimimus* for only a relatively short time. Russian expeditions into Mongolia in the late 1970s and early 1980s were the first to discover fossils of it, and it was described and named only as recently as 1981. Since then, Chinese paleontologists have unearthed other fossils, but remains of *Avimimus* are still very rare. Only three partial skeletons— and no complete skeletons—have so far been discovered.

head and beak

165

Sinosauropteryx

SIGH-no-saw-ROP-ter-ix

*S*inosauropteryx caused a sensation when it was revealed to the world in 1996. Here was a clear and perfectly preserved skeleton of a small theropod dinosaur, but covering most of the body were impressions of an enigmatic fuzz, a coat of fine filaments up to 1½ inches (4 cm) long.

Skin impressions found for some dinosaurs reveal that they had scales. But the growing recognition that birds evolved from dinosaurs raised the possibility that some dinosaurs had feathers, or at least some kind of protofeather-like body covering that later evolved into the feathers of present-day birds. However, feathers are relatively soft and rarely fossilized.

FIELD NOTES

◼ Chinese lizard feather

🦖 3 feet (1 m)

◼ Earliest Cretaceous

◼ Liaoning, China

🏛 Not on display

A juvenile Sinosauropteryx shows "protofeathers," an epidermal feature that evolved before true feathers, rising from behind the head and on the neck and back.

It is likely that several birdlike dinosaurs may have had feathers or similar structures, but, as they were not preserved, we know nothing about them.

Then *Sinosauropteryx* was found in the fabulous Liaoning deposits in China. Here was the right kind of dinosaur (a small, advanced theropod) with much of its back, rump, and tail covered in a fine, filamentous fuzz that appeared to be a furlike coat made up of thousands of short, single strands. While lacking the complex structure of a feather, it was a more intricate body covering than simple reptilian scales.

The fuzzy coat seemed to form a downy layer that would have been perfect for trapping body heat and keeping the animal warm. This observation lends more weight to the ongoing debate that some dinosaurs were warm-blooded. It is also possible that the fuzz was used when displaying to attract mates.

In one of the specimens, located right where the gut would have been in life, was the remains of its last meal, a small, unidentified mammal— providing proof that this little ancestor of the birds ate the ancestors of modern mammals. One of the other specimens revealed unlaid eggs in the oviducts.

Sinornithosaurus

SIE-nor-nith-oh-SAW-rus

The small predatory dino-saur *Sinornithosaurus* was a dromaeosaurid—one of a group of agile bipedal runners with large eyes, relatively large brains, and long, narrow snouts equipped with steak-knife teeth. The three fingers on each hand and the four toes of each foot had long, curved, wickedly sharp claws for hooking into prey, with the second toe claw being extra large. As in many other dinosaurs, especially fast-running kinds, the tail was stiffened by overlapping bony rods.

While much smaller than its famous relatives *Deinonychus* and *Velociraptor*, *Sinornithosaurus* would have been an equally effective predator on a small scale. It may have been a solitary hunter of large insects, lizards, primitive birds, and small mammals, but it is also possible that groups of *Sinornithosaurus* cooperated to hunt and bring down small plant-eating dinosaurs and other prey bigger than themselves.

The only known skeleton of *Sinornitho-saurus* was described in 1999 and is missing most of the vertebral

FIELD NOTES

- Chinese bird lizard
- 3 feet 6 inches (1.1 m)
- Early Cretaceous
- Liaoning, China
- Not on display

column and rib cage, but the skull, shoulders, hips, and limb bones are well preserved. Many details of the skeleton are similar to those of the primitive bird *Archaeopteryx*, providing evidence that dromaeosaurids were the dinosaurs most closely related to birds. The forelimbs are about four-fifths as long as the hindlimbs, and the shoulder sockets face to the side, allowing a wide range of movement, as in *Archaeopteryx*. Although unable to fly, *Sinorni-thosaurus* might have been able to "flap" its arms and snatch small flying prey out of the air.

Another birdlike feature of *Sinornithosaurus* is a dense covering of downlike filaments over most of its body. Many other dinosaurs may have had these protofeathers, but only under exceptional geological conditions are they preserved during fossilization. Similar structures are found in four other kinds of small theropods from the same locality in Liao-ning province, China.

Sinornithosaurus has several birdlike features, including a furcula, or wishbone, which all birds possess.

167

Mononykus

mon-oh-NIE-kus

M ost small theropods had long forelimbs and long, grasping hands. *Mononykus* and its closest relatives—its fellow alvarezsaurids—are conspicuous exceptions to this rule. *Mononykus* had very gracile hindlimbs, but its forelimbs were extremely short and its hand was effectively reduced to a single digit—the stout, clawed thumb for which the animal is named. An almost imperceptible second digit is no more than a nubbin.

structure of arm and hand

FIELD NOTES

■ One claw

🦖 3 feet (1 m)

■ Late Cretaceous

■ Mongolia

🏛 American Museum of Natural History, New York, USA

Mononykus was one of the prizes collected in joint Mongolian–American expeditions to Mongolia undertaken in the mid-1990s. Other alvarezsaurids are known from the Cretaceous of North and South America. It is also possible that members of this family occurred worldwide.

Mononykus's forelimbs were too short to be able to reach its face, but they were powerfully built—the construction of the elbows suggests the existence of large extension muscles.

In some ways they resemble the forelimbs of digging mammals such as moles. What would a small theropod with graceful legs and a slender body have done with such arms? Might *Mononykus* have been a digging animal that used its strong, though very short, arms to rip open termite mounds? Perhaps it was, but the rest of the skeleton is absolutely unlike that of any known burrowing animal.

The scientists that discovered *Mononykus* originally thought they had discovered a very primitive bird. It had small teeth and a bony tail, but the breastbone had a small keel and the fibula did not reach to the ankle—features found in birds more advanced that *Archaeopteryx*. However, it is clear that *Mononykus* was incapable of flight. Some recent analyses suggest that *Mononykus* was a very close relative of birds. Others classify it as being closely related to the ornithomimids, such as *Gallimimus*. Although it seems likely that *Mononykus* will remain classified as a theropod, its distinctly birdlike features are undeniable. In fact, a recently discovered alvarezsaurid from Mongolia preserves thin fibers that scientists think may be primitive feathers around parts of its body.

Caudipteryx

kaw-DIP-tuh-rix

The discovery of the bizarre theropod *Caudipteryx* in 1998 was an exciting event for the science of paleontology. Scientists had long realized that if birds evolved from dinosaurs, somewhere along the line there ought to be a dinosaur with half a wing—a limb that was more than a simple dinosaur arm but not developed into a fully functional wing. *Caudipteryx* was just such an animal. Here was a long-legged, gracile theropod dinosaur with well-developed but short arms—and impressions of small feathered wings trailing behind its forearms.

How *Caudipteryx* used these "wings" is a puzzle. They were too small to have been used for flight. Perhaps they helped the animal catch insects, or they may have combined with the tuft of feathers on the tail to make a stunning sexual display. Whatever their function, we can now be certain that an animal with half a wing once existed.

The exquisite state of preservation of *Caudipteryx* and other fossils from the Liaoning

FIELD NOTES

- Tail feather
- 3 feet (1 m)
- Early Cretaceous
- Liaoning, China
- Not on display

deposits reveals a wealth of detail. Not only were there long feathers on *Caudipteryx*'s arms and tail, but shorter downy feathers covered most of the body. The head was small and rounded. Long, sharp teeth were restricted to the very front of the mouth and projected more forward than downward. Perhaps they were part of an early horny beak. In the gizzard region there was a collection of tiny pebbles, or gastroliths, that the animal swallowed to help it grind up food—yet another birdlike feature. *Caudipteryx* had a well-developed wishbone, a feature seen only in birds and the theropods.

Caudipteryx's long legs suggest that it was a speedy runner that probably made a living chasing after insects or other small animals.

Feathers extending from the arms of Caudipteryx *can be seen clearly in this fossil. Above the feathers, gastroliths are visible. These were small stones that helped to grind up the food that the animal ingested.*

169

Segnosaurus

SEG-noh-SAW-rus

Few dinosaurs have caused as much speculation and debate among paleontologists as *Segnosaurus*. It has taken a long time to work out exactly what kind of dinosaur it was. *Segnosaurus* had a highly unusual collection of features that resembled bits of many other dinosaurs and dinosaur groups, combined with some uncommon characteristics only found in *Segnosaurus* and its close relatives, the segnosaurs.

The front of the snout of *Segnosaurus* was toothless and may have supported a beak, as in some ornithischian dinosaurs. *Segnosaurus* also had a hip arrangement similar to that of the ornithischians. The current consensus is that *Segnosaurus* and its close relatives *Erlikosaurus, Nanshiungosaurus,* and *Enigmosaurus* form a strange group of theropod dinosaurs. Features they share include a three-fingered hand; a four-toed foot; toes and fingers with curved claws; and a high, narrow skull. Curiously, the jaw curved downward and had rows of small pointed teeth along each side. This feature is also seen in the other segnosaurs.

FIELD NOTES

■ Slow lizard

🦖 19 feet 6 inches (6 m)

■ Late Cretaceous

■ Mongolia

🏛 Academy of Science, Ulan Baator, Mongolia

Exactly what use this strange combination of features was to *Segnosaurus* is widely debated. It has been suggested that it was a plant-eater descended from a meat-eating ancestor or, perhaps, a specialist termite hunter that used its huge claws to rip open termite nests. Alernatively, it may have been a specialist fish-hunting dinosaur, hooking fish out of the water with its claws.

Segnosaurus is a relatively new dinosaur, described in 1979, and known only from fragments and isolated bones. This makes it difficult to understand what it was really like.

Mongolia and China have produced many unusual theropods, such as *Segnosaurus*, from late Cretaceous deposits. These groups are not found anywhere else, indicating that what is now central Asia was isolated from the rest of the world by mountains and seas for most of the later Mesozoic era.

fossil remains of fish that may have been food for *Segnosaurus*

Saltasaurus

SALT-uh-SAW-rus

When *Saltasaurus* was described in 1980, it was quite a surprise for a number of reasons.

First, *Saltasaurus* was the first sauropod to be found with dermal armor. Previous fragments of armor had been found in the area and were thought to belong to an otherwise unknown ankylosaur. The armor comprised bony studs that interconnected to form a shield over the back of the animal. These varied from pea-sized to the size of a adult human's fist. Although the larger lumps tended to be aligned into rows, the shield did not exhibit any formal pattern.

The second surprise is that *Saltasaurus* comes from rock laid down almost at the very end of the age of the dinosaurs. Sauropods dominated the late

FIELD NOTES

■ Salta lizard

🦕 39 feet (12 m)

■ Late Cretaceous

■ Salta, Rionegro, Argentia; Palmitas, Uruguay

🏛 Argentine Museum of Natural Sciences, Buenos Aires, Argentina

skin detail

Jurassic but were scarce for most of the Cretaceous. Strangely, *Saltasaurus* and its kin, the titanosaurids, seem to have been making a reappearance just as the age of the dinosaurs was drawing to a close.

By sauropod standards, *Saltasaurus* was quite small. It was also quite stocky with relatively short, stumpy legs. The tail was long and ended in a whiplike lash similar to that of *Diplodocus*. The bones of the tail interlocked, thus stiffening the whole structure and possibly providing support for the animal when it reared up on its hind legs.

Several specimens of *Saltasaurus* have been found, representing three species. Most of the skeleton is known except for the greater part of the skull and many of the foot bones.

Saltasaurus was one of the dinosaurs named by the famous Argentinean paleontologist José Bonaparte. Bonaparte has described many dinosaurs from Argentina, including the strange horned theropod *Carnotaurus*.

eating from tree tops

Edmontonia

ed-mon-TOH-nee-ah

Edmontonia was one of the largest nodosaurids, one of the two main groups of the armored ankylosaurs. Nodosaurids characteristically had a boxlike head and bony armor covering the neck, back, and upper surfaces of the tail. This armor consisted of three types of bony elements embedded in the skin. The largest were pronounced spikes, on the shoulders and forming two rows running along the sides of the animal. Shieldlike scutes of varying sizes were arranged in several rows, running lengthwise from the back of the neck to the tip of the tail. In between the scutes and spikes were thousands of small, pea-sized ossicles. Together the spikes, scutes, and ossicles formed an impenetrable shield against the attacks of predators. Even the head had a set of interlocking bony plates over the upper surfaces to protect the brain, eyes, and nose.

The huge spikes on the shoulders gave *Edmontonia* an offensive weapon. By tucking its bony head below them, *Edmontonia* could drive these spikes forward into an attacker with potentially lethal effect.

FIELD NOTES

■ of Edmonton

🦖 23 feet (7 m)

■ Late Cretaceous

■ Alberta, Canada; Montana, Texas, USA

🏛 The Royal Tyrrell Museum of Paleontology, Drumheller, Alberta, Canada

detail of head with reinforced plates

As with most nodosaurids, *Edmontonia*'s belly was unprotected by armor and would have been vulnerable to attack if the animal were flipped over. To prevent this from happening, *Edmontonia* was very low-slung with relatively short, stumpy legs spread wide by broad hips and shoulder girdles. *Edmontonia* was built rather like a huge coffee table! *Edmontonia*'s boxy head had a cropping beak at the front of the mouth and rows of small, serrated, triangular teeth in the cheeks. Wear on these cheek teeth indicates that *Edmontonia* snipped its food into tiny pieces before swallowing it.

Edmontonia was one of the last of the nodosaurid dinosaurs, appearing late in the age of dinosaurs. Other nodosaurids are known from around the world in older rocks dating back to the latest Jurassic.

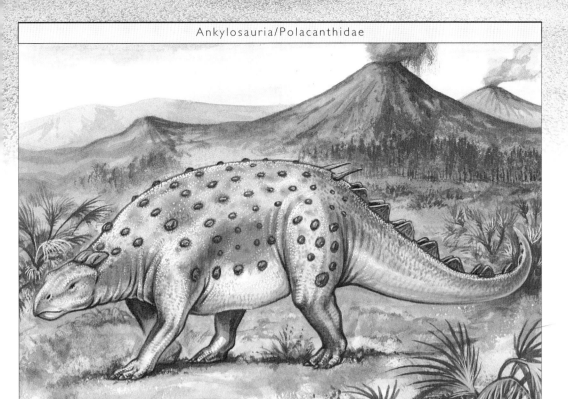

Minmi

MIN-mee

The most complete and best known dinosaur from Australia, *Minmi* was also the first armored dinosaur found in the Southern Hemisphere.

While it is clearly one of the armored ankylosaurian dinosaurs, it is not clear whether *Minmi* is a nodosaurid, an ankylosaurid, or a representative of a new, third group. Like other ankylosaurs, *Minmi* had a back covered by rows of bony shields and nubbins. Unlike most other ankylosaurs, however, this armor extended onto the flanks and belly and even onto the upper parts of all four legs. The tail had two rows of large, bladed spikes occurring in pairs for most of its length.

Similar to other ankylosaurs, *Minmi* was a broad but squat animal with four short legs. The hips were very wide and formed a bony raft across the rump. The head was generally flattened and quite broad at the back, narrowing to a thin snout. The front of the snout had a horny beak for snipping off leaves. The leaves were then sliced into small pieces by rows of small teeth along the sides of the mouth.

Minmi was first described from a set of three vertebrae found in southern Queensland in the 1960s. In 1990 an almost complete skeleton was found near Hughenden in central Queensland. The prized skeleton is in an extremely hard limestone that is slowly being dissolved in weak acid. It will take some years before the fossil's secrets are revealed.

Australia in the early Cretaceous was still connected to Antartica and was much farther south than it is today, with only its northernmost parts outside what is now the Antarctic Circle. The fact that Australian dinosaurs such as *Minmi* are so different from dinosaurs occurring elsewhere in the world at that time indicates that the emerging continent had only very limited access for land animals.

FIELD NOTES

- Named after Minmi Crossing, where it was first found
- 8 feet (2.5 m)
- Early Cretaceous
- Queensland, Australia
- The Queensland Museum, Brisbane, Australia

This skeleton of Minmi was found in Queensland, Australia, in 1990.

Euoplocephalus

yoo-oh-ploh-SEF-uh-lus

Built like an armored tank, *Euoplocephalus* ambled through the late Cretaceous landscape, well equipped to withstand attack from any other dinosaur. Low slung and broad, the back of *Euoplocephalus* bore rows of bony shields with some taller spikes over the shoulders and at the base of the tail. There were also spikes on the dinosaur's cheeks and behind each eye, protecting the head.

The most lethal weapon in *Euoplocephalus*'s armory was the double-headed club at the end of the long, stiffened tail. The base of the tail was quite flexible, but the last third was welded into a stiff rod by long struts growing out of each vertebra. The tail club could be swung most effectively from side to side, swiping at the feet of an attacking predator. If it connected with full force, it could shatter the ankle bones of the attacker, a wound that could later prove fatal.

Euoplocephalus had a compact, rounded head. Like other ankylosaurids, but unlike nodosaurids, it had

FIELD NOTES

- Well-armored head
- 🦖 23 feet (7 m)
- Late Cretaceous
- Alberta, Canada; Montana, USA
- 🏛 National Museum of Natural Sciences, Ottawa, Canada

a complex and convoluted nasal passage in the skull, but the function that this served is not clear. Perhaps the extra length given by the twists and loops allowed air to be warmed while the animal was breathing in, or perhaps this passage collected moisture from air being exhaled. The passage may also have been lined with sensors that gave *Euoplocephalus* an enhanced sense of smell for detecting food, predators, or potential mates.

The mouth had a broad beak at the front and a wide palate lined with small teeth. This arrangement suggests that *Euoplocephalus* was not particularly selective about what it ate and would consume almost any plant material that it could reach.

Around 40 specimens of *Euoplocephalus* have been found. All were isolated finds, which suggests that these animals were loners rather than pack or herd animals. Packs and herds provided plant-eaters with a defense against predators but, perhaps because it was so heavily armored, *Euoplocephalus* had no need to rely on group behavior for protection.

A Euoplocephalus skull

Ankylosaurus

an-KEE-loh-SAW-rus

FIELD NOTES

- Stiff lizard
- 33 feet (10 m)
- Late Cretaceous
- Montana, Wyoming, USA; Alberta, Canada
- Provincial Museum of Alberta, Edmonton, Canada; American Museum of Natural History, New York, USA

The last and largest of all the armored dinosaurs, *Ankylosaurus* was found in some of the very youngest beds that contain dinosaur fossils. With its massive tail club and a suit of armor, a solitary *Ankylosaurus* would have had nothing to fear from most predators.

Ankylosaurus gives its name to the group to which it belongs (the ankylosaurs) and takes its name from the bony nature of its skeleton. "Ankylosed" means "stiffened with bone"—and this is a fair description. The head was covered with an extra layer of bony plates and spikes, the back and tail were covered by interlocking bony shields, and the vertebrae at the end of the tail were welded together by bone.

The heavy, bony tail club was wielded like a wrecking ball and could do serious damage to any attacking predator. Despite the weight of the tail club and its position at the end of a long, heavy tail, the club was carried clear of the ground. Trackways made by *Ankylosaurus* and its kin show no signs of a tail dragging behind.

Like its close relative *Euoplocephalus*, *Ankylosaurus* had a complex systems of loops and twists in its nasal passages. These may have been for warming air, reclaiming water from expiring air, or for enhancing the sense of smell. It is also possible that they were used as a resonating chamber, helping the creature to make loud mating or distress calls.

Ankylosaurus was one of the many dinosaurs found and named by the famous paleontologist Barnum Brown. In 1910, Brown found a particular specimen that is now on display in the American Museum of Natural History in New York. As the story goes, he didn't have time to dig the specimen out himself, so he paid some local ranchers to do the job for him. When he returned a year later, the ranchers had excavated nearly 1,180 cubic yards (900 m³) of solid sandstone, mostly by hand but also using a little dynamite. Despite this stupendous effort, only a partial specimen was recovered.

tail club

175

Hypsilophodon

HIP-sill-OFF-oh-don

One of the most famous of small dinosaurs, *Hypsilophodon* was also one of the earliest to be studied. It was discovered in 1849, in the same Wealden rocks in southern England that had yielded fossils of the ornithopod *Iguanodon* 20 years earlier. Indeed, at first *Hypsilophodon* was thought to have been a juvenile *Iguanodon*. In 1869, the English zoologist Thomas Henry Huxley recognized that it was a different animal and named it after the strong ridges that were visible on its teeth. Huxley realized that this small animal would have been agile and would have moved principally on its hind legs. This was a suprising view, considering that at the time most scientists considered dinosaurs to be ponderous creatures that moved about on all fours.

Early reconstructions of *Hypsilophodon*'s foot mistakenly showed it to have a reversed hallux (first toe). Such a reversal is common in animals that live in trees, as it allows them to grip branches while they perch. As a result, early

FIELD NOTES

■ High-ridged tooth

🦕 7 feet (2.1 m)

■ Early Cretaceous

■ Isle of Wight, England

🏛 Natural History Museum, London, UK

skull

reconstructions of a complete *Hypsilophodon* showed it perched up in branches, rather like an ancestral tree kangaroo.

Eventually the foot was reconstructed correctly and *Hypsilophodon* was brought to ground as a fast-running ornithopod, similar in many respects to the Jurassic *Dryosaurus*. Like *Dryosaurus*, *Hypsilophodon* had long hind legs that were well suited for running at high speed. *Hypsilophodon*'s tail, like *Dryosaurus*'s, was long and was held stiffly off the ground to counterbalance the weight of its body.

Hypsilophodon's teeth and jaws were well adapted for grinding the tough plant matter on which it fed. It had a fleshy cheek, where food could be stored before it was chewed, and a horny beak that cropped food as it entered the mouth. Unlike *Dryosaurus,* and most other ornithopods, *Hypsilophodon* still retained some front teeth in the upper jaw. In this respect, *Hypsilophodon* was probably more primitive than *Dryosaurus*, even though it lived 300 million years after the more advanced ornithopod had become extinct.

Leaellynasaura

lay-ELL-lye-nuh-SAW-ruh

Since 1978, hundreds of dinosaur bones have been collected from early Cretaceous rocks at a site known as Dinosaur Cove in Victoria, Australia. There are very few complete skeletons in these rocks—the bones are from many different animals and they have been jumbled together—but studies have shown that most of the bones belonged to small dinosaurs that were closely related to *Hypsilophodon*. The first of these new "hypsilophodontids" to be named by science was *Leaellynasaura*.

The material that has so far been found relating to *Leaellynasaura* is incomplete, but it does indicate that this was a small dinosaur, even in comparison with other hypsilophodontids.

FIELD NOTES

■ Leaellyn's lizard

🦕 3 feet (1 m)

■ Earliest Cretaceous

■ Victoria, Australia

🏛 Museum Victoria, Melbourne, Australia

The first *Leaellynasaura* bones are from an animal that would have weighed no more than about 2 pounds (0.9 k) and was almost certainly a juvenile. One specimen preserves an internal cast of the braincase. It shows that the optic lobes (the parts of the brain that process visual information from the eyes) were very large compared with those of other dinosaurs. As with other hypsilophodontids, the bone tissues show that *Leaellynasaura* grew continuously and quickly. Seasonal variations in temperature did not slow down its growth rates, and this has led some scientists to speculate that this dinosaur may have been warm-blooded.

During the early Cretaceous, Victoria was well within the Antarctic polar circle. This means that *Leaellynasaura* was living, and apparently thriving, at latitudes that no reptile lives at today. The fact that even juveniles had enlarged optic lobes suggests that this dinosaur had large eyes that helped it to see its way through the long, dark polar winters.

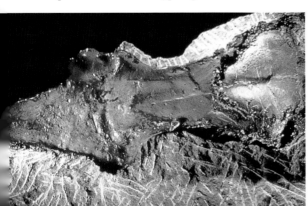

This fossilized Leaellynasaura skull, from Dinosaur Cove in Victoria, Australia, is still partly encased in rock.

177

Qantassaurus

KWON-tuh-SAW-rus

Named for Qantas, the national airline of Australia, *Qantassaurus* was a bipedal herbivore. It had a short, deep head, five-fingered hands, long legs with four-toed feet, and a tail roughly as long as the head and body combined. It was about the same size as a modern kangaroo, but it was a running animal rather than a hopper—or so we can presume, for this description applies to other hypsilopho-dontids known from more complete remains. Hypsilophodontids, the group that was close to the distant ancestors of later ornithopods, first appeared in the middle Jurassic, but they became surprisingly diverse in the early Cretaceous of what is now south-eastern Australia.

Lower jawbone of Qantassaurus

Since 1978, many isolated or broken bones and teeth of hypsilophodontids have been found in the channel and floodplain sediments in the Otway and Strzelecki ranges on the south coast

FIELD NOTES

◼ *Qantas lizard*

🦖 6 feet (2 m)

◼ *Early Cretaceous*

◼ *Victoria, Australia*

🏛 *Museum Victoria, Melbourne, Australia*

of Victoria. These sediments were formed in the widening rift valley between Australia and Antarctica, at a time when the whole area lay within the south polar circle. At one locality that is known as Dinosaur Cove, a tunnel has been cut into the hard rock at the base of the seacliff to mine bones from an ancient stream channel. From the range of different-sized and different-shaped femora (upper leg bones), it has become clear that various kinds of hypsilophodontids—possibly five or six—coexisted in this habitat. Most other parts of the skeleton are less helpful, but fortunately the high-crowned, many-ridged teeth of hypsilophodontids can be used to distinguish different species.

In 1999, *Qantassaurus* became the fourth of these species to be named, but the only specimens that paleontologists believe definitely belong to it are remains of lower jaws and teeth. Because *Qantassaurus* shared its habitat with species of *Fulgurotherium*, *Atlascopcosaurus*, and *Leaellynasaura*, it will be difficult to say which other bones belonged to each of these somewhat similar dinosaurs.

Atlascopcosaurus

AT-las-KOP-koh-SAW-rus

Atlascopcosaurus was one of several small hypsilopho-dontid dinosaurs that lived in south-eastern Australia during the early Cretaceous, at a time when the rift between Australia and Antarctica was just beginning to open. This dinosaur was named for the company that provided rock drills and compressors for the Dinosaur Cove digs in Victoria in the late 1980s. Dr. Tom Rich and Dr. Patricia Vickers-Rich, who conducted these digs, described *Atlascopcosaurus* in 1989.

Like other hypsilophodontids, *Atlascopcosaurus* was an agile bipedal plant-eater. It played an ecological role similar to that of forest antelopes or wallabies in the modern world. Its upper teeth were similar to those of *Zephyrosaurus*, from Montana, USA, but the primary ridge on each of *Altascopcosaurus*'s teeth was more strongly developed. *Zephyrosaurus* was also somewhat smaller. *Atlascopcosaurus* used its high-crowned, many-ridged teeth for browsing on the tough ferns and horsetails that formed the understory of the forests along the rift valley.

Like its slightly smaller relatives *Leaellynasaura*, *Fulgurotherium*, and *Qantassaurus*, *Atlascopcosaurus* probably lived in family groups or small herds,

FIELD NOTES

- Atlas Copco lizard
- 9 feet (2.7 m)
- Early Cretaceous
- Victoria, Australia
- Museum Victoria, Melbourne, Australia

moving between resting places in deep thickets and open clearings, where it fed on new plant growth. With its long legs and almost hooflike hind claws, this dinosaur could outrun most of its potential predators.

An articulated partial skeleton thought to be that of *Atlascopcosaurus* is remarkable because the left tibia (shin bone) shows severe pathology—evidence that the animal suffered from chronic osteomyelitis for the last few years of its life. The fact that a crippled animal was able to survive for several years suggests that *Atlascopcosaurus* was not always under intense pressure from predators, and also that it probably did not need to undertake long migrations to avoid the harsh polar winters.

The Dinosaur Cove mine has yielded several new species.

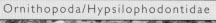

Tenontosaurus

ten-ON-toh-SAW-rus

FIELD NOTES

- Tendon lizard
- 26 feet (8 m)
- Early Cretaceous
- Montana, Wyoming, USA
- Peabody Museum of Natural History, New Haven, USA; Academy of Natural Sciences, Philadelphia, USA

Most ornithopod dinosaurs have strong tendons running along their backbone in the hip region. On *Tenontosaurus* these tendons were so well developed that they inspired the dinosaur's name. These thick tendons were ossified (strengthened with bone) and were arranged in bundles that ran parallel to the length of the spine. As a result, the spine of this dinosaur, from the lower back to the upper tail, would have been very stiff. This would have helped the animal to support the weight of its body across its hips; the torso was held stiffly in front of the hips and was balanced by the tail, which stretched out horizontally behind. Because the spinal column was almost parallel to the ground, the tail, which was very thick and bulky, would have been held high. This is confirmed by trackways of ornithopods which show no signs of marks made by tails dragging along the ground.

Tenontosaurus fed on plant matter, which it broke off with its horny beak and then chewed with its teeth, all of which were at the back of its mouth.

The small, swift-moving *Deinonychus* was *Tenontosaurus*'s main enemy. The teeth of this predator have been found along with *Tenontosaurus* skeletons. Some scientists therefore believe that *Deinonychus* hunted the larger animal in packs, but there is no real evidence to support this. In spite of what would seem to be the advantages of having clawed feet and a huge tail, *Tenontosaurus* would have been easy catch for packs of sharp-fanged *Deinonychus*.

There is still debate about the proper classification of this dinosaur—some maintain it was an advanced hypsilophodontid while others argue that it was an early iguanodontid. *Tenontosaurus* does appear to have been a transitory form between the smaller ornithopods, such as *Dryosaurus*, which arose in the Jurassic, and the larger ornithopods, such as *Iguanodon*, which arose in the Cretaceous.

Tenontosaurus being attacked by five Deinonychus

180

Ouranosaurus

oo-RAN-oh-SAW-rus

T his early Cretaceous orni-thopod from west Africa is one of the most puzzling dinosaurs ever discovered, and paleontologists are still trying to work out what it looked like. This is surprising, because *Ouranosaurus* is known from an almost complete skeleton—which was discovered in 1966—and its closest known relative, *Iguanodon*, is one of the best understood dinosaurs of all.

The problem lies with the backbone. The neural spines—the bones that projected upward from the main part of the vertebrae and which usually supported important sets of the back muscles and tendons—were simply huge in *Ouranosaurus*. Imagine an animal with a chest the size of a modern racehorse's, with a set of spines more than 27 inches (0.7 m) tall along the length of its back. What function could these spines have served?

Some scientists believe that *Ouranosaurus* sported a huge sail-like structure on its back. "Sail-backed" animals are known from the Permian period,

FIELD NOTES

■ Brave lizard

🦖 23 feet (7 m)

■ Early Cretaceous

■ Niger

🏛 National Museum of Niger, Naimey, Niger; Civic Museum of Natural History, Venice, Italy

260 million years ago, and it is thought that these "sails" helped them to regulate their body temperature. In some recon-structions, then, *Ouranosaurus* is shown as an unremarkable, medium-sized ornithopod, except for its long sail.

Others disagree with the "sail" theory. They point out that in the hot, dry climate in which it lived, *Ouranosaurus* would not have needed a sail to get warm. Overheating would have been more of a problem, and a sail would have been of no use in getting rid of heat. Plus, the spines of *Ouranosaurus* bear little resemblance to those of the Permian sail-backs. They look more like the spines that form the withers in modern mammals such as bison. Some scientists therefore reconstruct *Ouranosaurus* as an unusual ornithopod with a huge humped back—a kind of dinosaurian version of a camel or buffalo.

skeleton showing bones that supported the hump on the back

181

Iguanodon

ig-WAHN-oh-don

I n a way *Iguanodon* could be called a "founding father" of dinosaurs—not in the sense that it is ancestral to all other dinosaurs, but in terms of our scientific understanding of dinosaurs. In 1825, *Iguanodon* became the second dinosaur to be named by science, and it was one of the three animals around which the British paleontologist Sir Richard Owen constructed the scientific concept of "Dinosauria" in 1842. Ever since, studies and reconstructions of *Iguanodon* have played an important role in advancing our knowledge of dinosaurs in general.

Iguanodon was widespread in the early Cretaceous—species have been described from Europe, North America, and Mongolia. The first fossils of this dinosaur to be found came from the Wealden rocks of southern England—rocks that were formed in an extensive series of shallow lakes and estuaries. The very first fossils were only teeth, which looked much like the teeth of living iguanas—hence the name. As more bones came to light, researchers reconstructed this dinosaur as a large quadrupedal herbivore—

FIELD NOTES

- Iguana tooth
- 33 feet (10 m)
- Late Jurassic
- Western USA; Western Europe; Romania; Mongolia
- Royal National Institute of Natural Sciences, Brussels, Belgium

hand showing thumb spike

a sort of reptilian rhinoceros— and, in what turned out to be one of the most celebrated mistakes in the history of paleontology, a large bony spike that had been found with other parts of the skeleton was placed on the end of the nose. This early reconstruction helped to reinforce the perception of dinosaurs as lumbering animals—a perception we now know to be inaccurate.

It took one of the most remarkable fossil finds of all time to shake this stereotype. In 1878, workers in a coal mine in southern Belgium unearthed a large bone full of what they thought to be gold. The "gold" turned out to be "fool's gold" (pyrite), but the possibility of riches inspired further digging and led to the discovery of dinosaur skeletons. After three years, the complete skeletons of 31 *Iguanodon* had been recovered—at the time, they were the best preserved dinosaur fossils

that had ever been found. Study of these specimens showed that, far from being a heavy, lumbering quadruped, *Iguanodon* was relatively light for its great length and could move on its back legs. These skeletons also put the nose spike where it belonged. In fact, *Iguanodon* had two spikes—one on each thumb.

For almost a century after that, *Iguanodon*, and other large ornithopods, such as the hadrosaurids, were reconstructed rather like colossal kangaroos, standing on their hind legs with the head held high and the tail stretching out along the ground. It was not until the 1970s that new studies revealed that *Iguanodon* had strong front legs and that the central three fingers ended in hooves. This implied that

A reconstructed Iguanodon *skeleton, based on an original in the Royal National Institute of Natural Sciences in Brussels, Belgium.*

Iguanodon spent some time on all fours, and probably ran on its hind legs when it needed to move quickly. As in other ornithopods, its spine was supported by large, ossified tendons around the hips; unlike in other ornithopods, however, the tendon bundles were arranged diagonally in a trellis pattern, rather than parallel to each other.

The jaws and teeth of *Iguanodon* made it an efficient plant eater.

The roughened bulge halfway down this Iguanodon *hip bone shows where it broke and healed slightly out of alignment.*

A formidable battery of closely packed cheek teeth were well suited to grinding up tough plant matter—the upper surface of each tooth was broad and ridged. The jaw bones that held the teeth moved upward and outward in the skull as the animal chewed, thus allowing the grinding surfaces to move against each other and thereby contributing to the efficiency of the chewing action. The same arrangement has been observed in a number of other ornithopods, but in *Iguanodon* it appears to have been particularly well developed.

Edmontosaurus

ed-MONT-oh-SAW-rus

The hadrosaurids, or duck-billed dinosaurs, were a family of large ornithopods, which are informally named after the wide, flattened front part of their mouths. This was covered in a horny, toothless beak and looked like the bill of a monstrous duck. *Edmontosaurus* would have used this beak to bite off plant matter, which it chewed with the many tightly packed teeth—there may have been up to a thousand— that lined both of its jaws. As teeth wore out, they were replaced with new ones.

Edmontosaurus was one of the last hadrosaurids, and also one of the largest. In fact, it was, along with *Tyrannosaurus* and *Triceratops*, one of the last surviving dinosaurs, living right to the end of the Cretaceous. To help it cope with its huge body weight—a large *Edmontosaurus* may have weighed up to 5 tons (5.1 t)—it had

FIELD NOTES

■ Edmonton lizard

🦕 42 feet (13 m)

■ Late Cretaceous

■ Alberta, Canada

🏛 Peabody Museum of Natural History, New Haven, USA; Smithsonian Institution, Washington DC, USA; Royal Ontario Museum, Toronto, Canada

strong front legs and hooves on its "hands." As well, its spine was supported by huge bony tendons, which criss-crossed all the way down. *Edmontosaurus* probably spent most of its time on all fours, rising up on its back legs only when it needed to run.

Spectacular finds of *Edmontosaurus* from Alberta, Canada, have preserved impressions of the skin around parts of the body, including the "hand." It was one of these "mummified" fossils that led to early reconstructions of hadrosaurids as mainly aquatic animals. The skin of *Edmontosaurus*'s hand shows a structure between the fingers that looks like the webbed foot of a duck. It was realized only later that what looked like webbing was really the remains of padding behind the hooves. Many present-day hoofed animals have similar padding, which helps to bear the animals' weight. Modern reconstructions of *Edmontosaurus* and other hadrosaurids show them as fully terrestrial animals.

The strong jaws of Edmontosaurus contained tightly packed rows of tiny, leaf-shaped teeth that were ideal for grinding.

184

Maiasaura

MY-uh-SAW-rah

FIELD NOTES

■ Good mother lizard

🦕 30 feet (9 m)

■ Late Cretaceous

■ Montana, USA

🏛 Museum of the Rockies, Bozeman, USA

protecting the nest

Ornithopod nest sites have provided scientists with excellent opportunities to see how dinosaurs were born and grew up. It seems that hypsilophodontid babies could walk as soon as they were hatched, and they may have left the nest soon after. Hadrosaurids had a rather different reproductive strategy. No animal shows that better than *Maiasaura*.

In fact, the first fossils ever found of *Maiasaura* were a huge nesting colony, about 75 million years old. It was discovered in the badlands of Montana in 1978 by John Horner and Robert Makela. This colony contained eggs (many of them still intact), babies, and adults; even the arrangement of the eggs in the nest could be

The reconstructed skeleton of a juvenile Maiasaura. *As these animals grew, their heads became flatter and wider.*

seen. Careful study of the site led to some interesting insights into the nurturing habits of *Maiasaura*. Many of the baby *Maiasaura* were clearly too large to be newly hatched but were evidently still living in the nest. Like the leg bones of some species of modern birds, the bones in the legs of the baby *Maiasaura* were not fully formed. Despite this, their teeth showed signs of wear. The logical conclusion was that the babies were being fed in the nest. This seeming demonstration of parental care inspired the name of this dinosaur, which was bestowed on it by its discoverers in 1979.

It appears that young *Maiasaura* grew quickly. To some researchers, this suggests that they were warm-blooded. The nests that Horner and Makela found also throw light on the social organization of these hadrosaurs. The number and proximity of the nests indicate that females nested in large groups. Some scientists believe that *Maiasaura* were strongly social animals that lived in herds of many thousands.

Corythosaurus

koh-RITH-oh-SAW-rus

The best known of the crested duckbill dinosaurs (lambeosaurine hadrosaurs), which lived beside the ancient inland sea of western North America, *Corythosaurus* walked on all four limbs. It had flattened, blunt claws on its four-fingered hands but, as with other ornithopods, most of its body weight was supported by the large three-toed hindlimbs and balanced by the large tail. Criss-crossing ossified tendons stiffened the tail all the way from the hips, preventing the tail from swinging from side to side when the dinosaur ran. The spine was strongly flexed, or "hunched," at the shoulders, suggesting that *Corythosaurus* fed on low-growing plants—probably on the flowering plants that had evolved earlier in the Cretaceous—but that it could also raise its head above shoulder level to check for danger and to communicate with other members of the herd.

Like other duckbills and some other late Cretaceous plant-eating dinosaurs, *Corythosaurus* had huge numbers of teeth crammed together

FIELD NOTES

- Corinthian-crested lizard
- 33 feet (10 m)
- Late Cretaceous
- Alberta, Canada; Montana, USA
- American Museum of Natural History, New York, USA; Royal Ontario Museum, Toronto, Canada

skull showing crest

into "batteries" forming a single grinding surface on each side of the upper and lower jaws. This allowed the dinosaur to process large amounts of food at once. The hadrosaurines had broad, "ducklike" snouts to cut a wide swathe through the herb layer, while lambeosaurines such as *Corythosaurus* had narrower snouts and presumably fed more selectively.

The most distinctive feature of the lambeosaurines was the hollow bony crest on top of the head. The size and shape of these crests varied greatly. As a result, different skeletons of *Corythosaurus* have been identified as belonging to at least seven different species. However, comparison of more than 20 skulls has shown that the crest changes as it grows and differs between the sexes. Only a single species is, therefore, now recognized. The large-crested individuals are thought to be the adult males. They probably used the crest in behavioral displays to attract mates and to intimidate other males. The skin covering the crest may have been brightly colored or patterned, and the hollow within the bone, which was connected to the airway, may have been used to produce distinctive honking calls.

Lambeosaurus

LAM-bee-oh-SAW-rus

One of the largest of the crested duckbill dinosaurs (lambeosaurine hadrosaurs), *Lambeosaurus* lived in the same area and at the same time as several other members of this group of low-browsing herbivores. It seems, in fact, that several species of *Lambeosaurus* lived at the same time. They were distinguished by different-shaped bony crests on the tops of their heads, in much the same way that different kinds of modern deer and antelope have different-shaped antlers or horns. *Lambeosaurus lambei* had a hatchet-shaped crest projecting slightly forward from the top of its skull and a solid spur farther back on the head, whereas *L. magnicristatus* had a single-piece crest more like that of *Corythosaurus*. As in other lambeosaurines, the hollow crest

FIELD NOTES

■ Lambe's lizard

🦖 49 feet (15 m)

■ Late Cretaceous

■ Alberta, Canada; Montana, USA; Baja California, Mexico

🏛 Royal Tyrrell Museum, Alberta, Canada

would have formed a resonating chamber for its calls, amplifying them and making a distinctive sound in each species. The shape and patterning of the crest would also have helped individuals to recognize each other in the herd.

Nearly 20 skulls and skeletons of *Lambeosaurus* have been described. One fossil deposit, a "bonebed" containing hundreds of jumbled up skeletons buried by floods, includes specimens of *Lambeosaurus* along with *Corythosaurus*, *Prosaurolophus*, *Gryposaurus*, and *Parasaurolophus*, which suggests that these duckbills shared the same habitat and may even have migrated in huge mixed herds.

Some *Lambeosaurus* fossils display detailed impressions of the skin, showing that the skin of the body had a "pebbly" texture and that a weblike sheath of skin joined the fingers. When they were first described, these "webbed hands" were thought to prove the now-outmoded idea that duckbills were aquatic. The "web" actually enclosed a fleshy pad on the palm like that on a camel's foot.

Barnum Brown built a boat to reach the rich fossil sites along the Red Deer River, in Alberta, Canada..

187

Saurolophus

SAW-roh-LOW-fus

The most distinctive feature of *Saurolophus*—the one that gives this dinosaur its name—was the sharp, pointed ridge of bone that projected from the top of its head. This large hadrosaur is now known from several well-preserved, complete skeletons. The first species to be discovered, *Saurolophus osborni*, was named by Barnum Brown in 1921. It was based on a complete skeleton and additional skull material collected from the Horseshoe Canyon Formation of southern Alberta, Canada. A second species, *S. angustirostris*, was named in 1952 by the Russian paleontologist A. Rozhdestvensky. This species, which was the larger of the two, and which also had a larger crest, is one of the most common dinosaurs to be found in the latest Cretaceous beds of the Gobi Desert in Mongolia.

Some specimens of *Saurolophus* are so well preserved that they show skin impressions. From these we can tell that *Saurolophus* had leathery, fine-scaled skin. Except for some minor

FIELD NOTES

- Ridged lizard
- 42 feet (13 m)
- Late Cretaceous
- Alberta, Canada; Mongolia
- American Museum of Natural History, New York, USA; Paleontological Institute, Moscow, Russia

Saurolophus may have inflated the skin that covered its crest in order to make sounds or as a form of courtship display.

differences in the overall size and in the shape and height of the crest, the two species of *Saurolophus* are almost identical. This strongly suggests that by the end of the Cretaceous this dinosaur had a widespread distribution across the Northern Hemisphere.

The pointed ridge on *Saurolophus*'s head may have been covered by fleshy nostrils or nostril flaps. *Saurolophus* may have used its bony skull structure to send honking-like sound signals to other members of its species, perhaps, as a form of courtship display. It is possible that colored skin covered the crest and stretched between it and the back of the animal's head; *Saurolophus* may have been able to inflate this skin covering by breathing through a hole in the front of the crest.

Saurolophus had large numbers of closely packed teeth that were well suited to chewing the hard plant material, such as ferns and conifers, that constituted its diet.

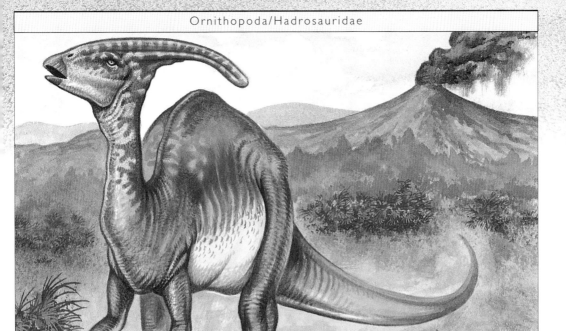

Parasaurolophus

PAR-uh-SAW-roh-LOH-fus

FIELD NOTES

■ Side-ridged lizard

🦖 35 feet (10.5 m)

■ Late Cretaceous

■ Montana, New Mexico, USA;
Alberta, Canada

🏛 Royal Ontario Museum,
Toronto, Canada; Los Angeles County
Museum, USA

With its snout bones drawn up into a giant snorkel-like structure, *Parasaurolophus* was one of the most bizarre of all the hadrosaurs. It lacked a hole in its apex, and because of this it is clear that this bony structure was not used as a breathing apparatus while the animal was swimming or feeding underwater. It seems more likely that it helped *Parasaurolophus* produce noises for signaling to mates or, if it was colored, for courtship displays. We know from the specimens that have been discovered that soft tissues adorned the bony crest.

The first skeleton of *Parasaurolophus* was collected in 1921 by Levi Sternberg in the region of southern Alberta, Canada, that is now the Dinosaur Provincial Park. This early find is still the most complete specimen to have been discovered. Three species of *Parasaurolophus* are recognized

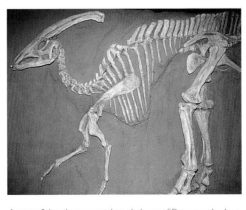

A cast of the almost complete skeleton of Parasaurolophus that Levi Sternberg found in 1921.

from their skulls. *P. cyrtocrystatus* had a short crest; on the other two species—*P. walkeri* and *P. tubicen*—the crests were much longer. The internal structure of the crest, which, unlike the crest of *Saurolophus*, had a hollow area that connected with the nostrils and the back of the throat, was more complex in *P. tubicen* than in other *Parasaurolophus* species.

Like all hadrosaurs, *Parasaurolophus* was a plant-eater. It had many closely compacted teeth, each of which had a central ridge. The teeth formed a strong dental battery that made it easier to chew tough vegetable matter

The principal dangers this dinosaur faced were from the larger predators such as *Albertosaurus*. *Parasaurolophus* probably sought protection from its enemies by living in large herds, in the same way that herbivores that inhabit the African plains do today.

189

Muttaburrasaurus

MUT-uh-BUH-ruh-SAW-rus

FIELD NOTES

■ Lizard from Muttaburra

🦖 33 feet (10 m

■ Early Cretaceous

■ Northern Queensland, New South Wales, Australia

🏛 Queensland Museum, Brisbane, Australia; Western Australian Museum, Perth, Australia

A large ornithopod that stood about 16 feet (5 m) high, *Muttaburrasaurus* is known from about 60 percent of its skeleton. A Mr. David Langdon found the skeleton in 1963 on Muttaburra Station in northern Queensland—hence this dinosaur's name. A second well-preserved skull, slightly older than the original specimen, was discovered on another property in north central Queensland. As well, a number of isolated bones and teeth of *Muttaburrasaurus* came to light on the Lightning Ridge opal field in northern New South Wales.

Muttaburrasaurus probably walked on all fours for most of the time. But it could also stand up on its hindlimbs to reach high into tree branches. Its most distinctive feature was a well-developed bump on the snout (called a "nasal bulla"), which it is thought may have housed an acoustic organ for calling to other dinosaurs.

Muttaburrasaurus had large areas of jaw muscle attachments, which greatly enhanced its chewing ability. This feature, in conjunction with teeth that were suited to shearing rather than to grinding food, has led some to believe that *Muttaburrasaurus* may have eaten meat from time to time.

Early reconstructions of the *Muttaburrasaurus* were modeled after *Iguanodon*, with a thumb spike. Scientists now believe this dinosaur was not closely related to the iguanodontids, but that it belonged to a family of its own. It may have been a large relative of the hypsilophodontid dinosaurs, such as *Atlascopcosaurus*. These dinosaurs inhabited the polar forests of Victoria and New South Wales in the early Cretaceous.

A skeleton of Muttaburrasaurus on display in the Queensland Museum, Australia.

Carcharodontosaurus

kuh-KAR-oh-dont-oh-SAW-rus

The very name of this dinosaur conjures powerful
images of how it may have
dispatched its prey—*Carcharodon*
is the generic name for the widely
feared ocean predator, the great
white shark. Like other theropods,
Carcharodontosaurus had teeth that
were serrated along the front and back.
However, *Carcharodontosaurus*'s teeth were
triangular and did not curve back as much
as those of most theropods. To Ernst Freiherr
Stromer von Reichenbach, the German paleontologist who first described this dinosaur in
the 1930s, these teeth seemed peculiarly
sharklike. This observed similarity led to the
naming of the dinosaur.

The remains on which Stromer had based
his description came from the late Cretaceous
of Egypt but they were far from complete. Although more fossils—including some claws and
teeth that French paleontologists found in the
Sahara in the 1970s—came to light elsewhere in
north Africa, all that could definitely be deduced
from them was that *Carcharodontosaurus* was a
huge creature. Eventually, during the 1990s,
Dr. Paul Sereno from the University of Chicago
discovered a skull of *Carcharodontosaurus* in
Morocco. When restored, this skull measured

FIELD NOTES

■ White-shark-toothed lizard

🦕 43 feet (13 m)

■ Late Cretaceous

■ Argentina

🏛 Not on display

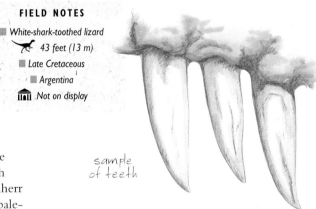

sample
of teeth

about 6 feet (2 m) long—as large as that of
Tyrannosaurus. However, the cavity that contained the brain of *Carcharodontosaurus* was
much smaller.

Carcharodontosaurus was obviously a close
relative of *Giganotosaurus* from South America.
Both of these dinosaurs had deep, domed snouts
and teeth that were smaller and more numerous
than those of *Tyrannosaurus*. It is possible that
these two contemporaneous, but geographically
separated, dinosaurs shared a common ancestor
that lived at a time when South America and
Africa were still parts of the same landmass. When
this landmass broke apart, different lineages could
have developed. Like *Giganotosaurus*, *Carcharodontosaurus* may have preyed on large sauropods. **191**

Pachycephalosaurus

PAK-ee-kef-AH-loh-SAW-rus

Incomplete dinosaur skulls featuring centrally thickened domes have been known from North America since the early years of the 20th century, but the nature of the group of dinosaurs to which these skulls belonged continued to be a mystery to paleontologists until 1940. In that year the first well-preserved complete skull—belonging to *Pachycephalosaurus*, the largest member of the group—was found in Montana. This skull had thick, spiky nodules of bone around the rim of the smoothly domed head and smaller protruding horns on the snout. *Pachycephalosaurus* is known only from its skull. No other fossils of this dinosaur have so far come to light.

Studies of the anatomy of *Pachycephalosaurus* skulls, and those of close relatives such as *Stygimoloch* and *Prenocephale*, indicate that these dinosaurs probably used their skulls in vigorous head-butting competitions during their

FIELD NOTES

■ Thick-headed lizard

🦖 26 feet (8 m)

■ Late Cretaceous

■ Wyoming, South Dakota, Montana, USA

🏛 American Museum of Natural History, New York, USA

A huge dome of bone protected the brain of Pachycephalosaurus *from powerful impacts, deflecting shock waves to the animal's backbone.*

charging attack

courtship battles, in much the same way that modern sheep and goats do. Alternatively, they may have engaged in head-to-head or head-to-flank pushing battles. The central part of the head consisted of very thick bone that would have acted like a helmet, protecting the dinosaur's small brain by carrying any shock waves away from the area of impact, down the sides of the head, to the backbone. The numerous ossified ligaments that strengthened the backbone would then have dissipated the effects of the shock.

The skull, which was almost 2 feet (60 cm) long, was nearly 8 inches (20 cm) thick at the central part the dome.

Pachycephalosaurus had triangular teeth with coarse serrations along the edges for shredding tough plant matter.

Stygimoloch

STY-gee-MOH-lok

FIELD NOTES

- River of Hell devil
- 20 feet (6 m)
- Late Cretaceous
- Wyoming, Montana, USA
- Not on display

With well-developed horns and spikes protruding from the base of its domed skull and from its snout, *Stygimoloch*—its name refers to the Hell Creek site in Montana, where it was found—was more elaborately ornamented than most other pachycephalosaurs. What we know about this relatively elusive plant-eating dinosaur has been gleaned from only about five skull fragments and parts of the body skeleton. Like its larger cousin *Pachycephalosaurus*, with which it shared the late Cretaceous North American landscape, *Stygimoloch* walked upright and probably had small forelimbs and a long, stiff tail. Males may have used the horns at the base of the skull for locking heads with opponents in head-pushing contests for winning mates. The horns, which were not very strong, may, on the other hand, have been purely ornamental and employed only for courtship displays.

Because fossil remains of *Stygimoloch* are so scarce, our understanding of this dinosaur is still very limited. Those parts of the skull that scientists have so far been able to study show that holes present in the rear of the skull roofs of many pachycephalosaurs—known as the "temporal fenestrae"—had closed up in *Stygimoloch*. This indicates that *Stygimoloch*'s skull was more robust and suggests that this dinosaur, along with *Pachycephalosaurus*, was one of the more advanced of the pachycephalosaurs.

Stygimoloch lived in a lowland habitat, where the principal predators would have been large theropods such as *Tyrannosaurus*, *Albertosaurus*, and *Aublysodon*. As well as being on the lookout for these, *Stygimoloch* would also have needed to be wary of a much smaller, though agile and rapacious, predator that probably hunted in large packs—*Dromaeosaurus*.

skull of Stygimoloch showing horns surrounding the domed skull and on the snout

193

Psittacosaurus

sih-TAK-oh-saw-rus

*P*sittacosaurus was discovered in Outer Mongolia in 1922, in the early stages of the famous expeditions undertaken by the American Museum of Natural History between 1922 and 1925. Henry Osborn named it for the beaklike appearance of its face. It is known from a number of well-preserved skeletons, which represent about eight different species from Mongolia, southern Siberia, and northern China, as well as from some lower jaw fragments that were discovered in northern Thailand.

Psittacosaurus was one of the earliest dinosaurs to show the typical beaked face of the ceratopsian group. This beak, which was supported by a single median bone—called the "rostral bone"—is the one feature that distinguishes the ceratopsians from all other dinosaur groups.

Psittacosaurus was one of the smallest and most primitive of the ceratopsians. It lacked the

FIELD NOTES

■ Parrot lizard

🦖 4 feet (1.25 m)

■ Early Cretaceous

■ Mongolia; China; southern Siberia; Thailand

🏛 American Museum of Natural History, New York, USA; Paleontological Museum, Moscow, Russia; Academy of Science, Ulan Baatar, Mongolia; Department of Mineral Resources, Bangkok, Thailand

well-developed frill and horns that were typical of more advanced ceratopsians, yet, along with the hard keratinous beak, it had the characteristic skull shape of a ceratopsian. It also featured, in common with later ceratopsians, the high palate and the sharp, slicing teeth with self-sharpening edges that were well suited to nipping off and shredding hard plant matter.

Psittacosaurus's hindlimbs were longer—although only slightly—than its forelimbs, which suggests that it could have moved about in an upright position for short distances. It may have done so to avoid attacks from predators or to forage in low-hanging tree branches.

Some skeletons of Psittacosaurus contain fossils of gastroliths— stomach stones that helped the animal to break down plant matter inside its stomach.

Psittacosaurus lived in the early Cretaceous of Mongolia and other parts of Asia about 90 million years ago. It had relatively long forelimbs and large, grasping hands.

194

Protoceratops

PROH-toh-SAIR-uh-tops

O ne of the earliest of the neoceratopsian dinosaurs, *Protoceratops* had a well-developed frill that extended back from the face and over the neck. However, it lacked the horns of the more advanced members of the group, although some species featured a small bump on the snout that may have supported a keratinous horn, similar to that of a modern rhinoceros.

Protoceratops was a very common animal in the late Cretaceous lowland habitats of Mongolia. Fossilized remains of this dinosaur were among the most abundant fossils found on the American Museum of Natural History expeditions to Mongolia that Roy Chapman Andrews led between 1922 and 1925. The abundance of these fossils has led scientists to believe that *Protoceratops* was a highly social animal that lived in herds.

The American expeditions also discovered eggs and nests belonging to *Protoceratops*. These were the first dinosaur nests ever found, and the discovery was widely publicized. Another famous find was that of a *Protoceratops* skeleton interlocked with that of a *Velociraptor*. Whether

FIELD NOTES

■ First-horned face

🦕 10 feet (3 m)

■ Late Cretaceous

■ Mongolia; China; Canada

🏛 American Museum of Natural History, New York, USA; Paleontological Museum, Moscow, Russia; Academy of Science, Ulan Baatar, Mongolia

skull
showing frill

or not this find represents an actual act of predation, *Velociraptor*, along with the larger theropods such as *Tarbosaurus*, would almost certainly have been among the main predators on a small plant-eater such as *Protoceratops*.

Thanks to the large number of complete skulls of *Protoceratops* that have been found, scientists have been able to distinguish differences between males and females. In adult males, the frill was more erect and there was a more prominent bump on the snout. This suggests that males used the larger frill, as well as the more protuberant snout bump, as a device to attract females. The bump may also have been used in fights between rival males.

Protoceratops seems to have moved about on all fours. The size and weight of its head and jaws would probably have made a bipedal stance impossible. It would have fed mainly on low-growing plant matter, which it broke off with its beak and then chewed with the many teeth in the back of its mouth.

195

Centrosaurus

SENT-roh-saw-rus

*C*entrosaurus was one of the most abundant of the large browsing ceratopsians at the end of the Cretaceous. Because its fossil bones have been discovered concentrated in a thick layer, scientists have theorized that tens of thousands of animals, roaming in large herds, were killed in a flood. The layer of bones, which is widespread throughout the rich dinosaur deposits of Alberta, Canada, is now known as the *Centrosaurus* bone bed.

This dinosaur had a well-developed frill that was made lighter by large holes (known as "fenestrae"). In life the frill and the holes would have been covered with skin. Small tongues of bone hung downward from the top of the frill, and there was a large curved horn near the front of the snout. *Centrosaurus* was first described in 1876 by Edward Drinker Cope and given the name *Monoclonius*, meaning "single horn." However, as the name and the description were based only on the partial skull of a juvenile, many scientists doubt the validity of the name. The dinosaur was renamed *Centrosaurus apertus*

FIELD NOTES

■ Sharp-point lizard

🦖 20 feet (6 m)

■ Late Cretaceous

■ Alberta, Canada

🏛 Royal Tyrrell Museum, Alberta, Canada; American Museum of Natural History, New York, USA

by Lawrence Lambe in 1904. Several other species were subsequently named, but these are now recognized as variations (possibly male–female differences) within a single species. Many complete *Centrosaurus* skeletons have recently been unearthed, including those of juveniles at different stages of development. Some skin impressions have also been found.

Centrosaurus, which moved on all fours, had powerful front limbs that would have enhanced tha animal's speed and agility. A ball-and-socket joint in the neck would also have been useful in defense. It allowed *Centrosaurus* to turn its head swiftly and bring its sharp horn into play against large predators, such as *Tyrannosaurus*, that attacked from the rear.

Reconstructed skeleton of Albertosaurus *standing over* Centrosaurus *at the Royal Tyrrell Museum, Alberta, Canada.*

Styracosaurus

sty-RACK-oh-SAW-rus

A moderately large ceratopsian, *Styracosaurus* is known from several skeletons and skulls. In most respects these skulls were similar to those of its closest relative, *Centrosaurus*. They were characterized by an extensive bony frill with two large openings, or fenestrae, situated symmetrically on either side of the frill. As with *Centrosaurus*, these openings would have significantly reduced the weight of the frill. Again like *Centrosaurus*, *Styracosaurus* had a large pointed horn on its snout and a pair of small horns above the eyes. In *Styracosaurus*, however, the snout horn was straight, rather than curved, and the frill was fringed with numerous smaller, sharp, projecting horns.

Of the two dinosaurs, then, *Styracosarus* seems to have been better equipped for defense. However, there is some doubt about the effectiveness of the horns on the frill. As they extended back over the neck, they certainly would have afforded some protection against predators and would have looked formidably threatening when viewed from the front, but as they stuck out to the side, they would have been difficult to employ as stabbing weapons. It is possible that the elaborate horned frill, which would have been covered with skin,

FIELD NOTES

■ Spiked lizard
🦖 18 feet (5.5 m)
■ Late Cretaceous
■ Alberta, Canada; Montana, USA
🏛 American Museum of Natural History, New York, USA

served more as an accoutrement for courtship display than as a weapon of defense. There is no doubt, however, that *Styracoaurus* would have used its deadly snout horn to defend itself, or even to make pre-emptive strikes, against potential predators such as *Tyrannosaurus*.

Styracosaurus was one of the first horned dinosaurs to be discovered. Lawrence Lambe found the first skull along the exposures of the Belly River in Alberta, Canada, and in 1913 he formally named the dinosaur *Styracosaurus albertensis*. In 1930, Charles Gilmore named a second species *S. ovatus*. Today, however, scientists generally regard them as being male–female variants of the same species.

horned skull of
Styracosaurus

197

Triceratops

try-SAIR-uh-TOPS

Triceratops is one of the best known of all dinosaurs and was the largest of the ceratopsians. Its massive head bore a short frill of solid bone along with the three large horns for which it is named—one above each eye and a smaller one on the snout. Traces of blood vessels found in the frill and horn have suggested to some paleontologists that the frill may have served as a means of regulating the animal's body temperature. As with other ceratopsians, the frill would have been covered with skin and may also have been used during courtship display.

Triceratops, which moved on all fours, had a heavy, robust body— this was necessary to support the weight of its head—and a short tail. Its solidly built forelimbs were shorter than the hindlimbs and do not seem to have been made for fast movement.

Fragmentary remains of large horned dinosaurs had been found in North America since 1855, but it was not until 1889 that John Bell Hatcher, who was searching the area around

FIELD NOTES

■ Three-horned head

🦕 30 feet (9 m)

■ Late Cretaceous

■ Alberta, Saskatchewan, Canada; Colorado, Montana, South Dakota, Wyoming, USA

🏛 American Museum of Natural History, New York, USA; Smithsonian Institution, Washington DC, USA; National Museum of Natural Sciences, Ottawa, Canada; National Museum of Natural History, Paris, France; Senckenberg Nature Museum, Frankfurt, Germany; Birmingham Museum, UK; Natural History Museum, London, UK; Hunterian Museum, Glasgow, UK; Royal Scottish Museum, Edinburgh, UK

Niobrara County, Wyoming, happened upon the first complete skull. Othniel Charles Marsh studied the discovery and in 1899 bestowed upon it the evocative name *Triceratops horridus*. Over the next three years, Hatcher collected about 30 neoceratopsian skulls, most of them identified as belonging to *Triceratops*. Barnum Brown is also credited with collecting many *Triceratops* skulls between 10 and 20 years later.

Some 16 different species were eventually named, about half of them by Marsh. Many of these species identifications were based on isolated horns and variable features of the skull. However, recent research by Dr. Catherine Forster, from the State University of New York at Stony Brook, has reduced the number of *Triceratops* species to just two—*T. prorsus*, which had straight horns above the eyes and a well-developed nose horn; and Marsh's original *T. horridus*, which was considerably larger but had a much smaller nose horn and slightly curved eye horns.

Triceratops's lifestyle was probably quite similar to that of that lumbering present-day herbivore, the rhinoceros. Its many rows of closely packed grinding teeth suggest that it was a feeder on a range of coarse vegetation, such as conifers, ferns, and cycads, as well as on some of the flowering plants that first appeared in the late Cretaceous. It cropped this plant matter with its long, powerful, pointed, horny beak. Its jaw mechanism was adapted primarily for cutting. Large jaw muscles attached from the lower jaw up onto the frill and powered the shearing action of the jaws.

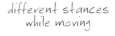

different stances while moving

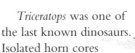

Triceratops was one of the last known dinosaurs. Isolated horn cores belonging to this dinosaur show that, along with its main predator, *Tryannosaurus*, it persisted right to the end of the Cretaceous. *Triceratops* may have charged at potential predators such as *Tyrannosaurus*, stabbing them with its three horns and using its thick, bony frill to protect its vulnerable neck. The great abundance of *Triceratops* fossils from the latest part of the Cretaceous provides convincing evidence of this dinosaur's ability to survive despite the numerous predatory theropods that shared its territory.

This skeleton of Triceratops horridus *shows its formidable armory of three horns and a bony neck plate.*

199

Chasmosaurus

KAZ-moh-SAW-rus

FIELD NOTES

■ Chasm lizard

🦖 17 feet (5 m)

■ Late Cretaceous

■ Texas, USA; Alberta, Canada

🏛 Royal Ontario Museum, Toronto, Canada

One of the earliest of the longer frilled dinosaurs, *Chasmosaurus* was a moderate-sized neoceratopsian. It had two short, upwardly curved horns above its eyes and a smaller horn on its snout. At the back of its long, narrow skull was a huge bony frill that stretched back over the animal's neck and shoulders. Within the frill were two enormous openings, or fenestrae, which made *Chasmosaurus*'s frill much lighter than that of any of its relatives. The bony part of the frill, which had right-angled upper corners with small ornamental horns, consisted of little more than a framework for the openings. This spectacular feature, covered with skin, was almost certainly a device for courtship display.

Lawrence Lambe, working with the Sternberg family on the Red Deer River in Alberta, Canada, discovered *Chasmosaurus*. He named it in 1914 after the chasm in which he found it. This dinosaur is now known from several skulls and skeletons. From the skeletons we can observe that the vertebral column above the pelvis was reinforced by many ossified ligaments. The pelvis itself was fortified by eight fused vertebrae. These helped dissipate the shocks that this large animal generated when it moved quickly.

Early studies made on skulls from Alberta and the northern United States suggested that they belonged to several different species. Today, however, most paleontologists consider that the variations reflect male–female differences within the same species— *C. canadensis*. Another species, *C. mariscalensis*, from Texas, is known from pieces of the skull and isolated bones which together make up an almost complete skeleton. *C. mariscalensis* differed from its northern counterpart mainly by having larger, more backward-curving horns.

stags challenging for dominance

Torosaurus

TAW-roh-SAW-rus

Torosaurus skull

human skull

FIELD NOTES

■ Bull lizard

🦖 25 feet (7.5 m)

■ Late Cretaceous

■ Wyoming, South Dakota,
Colorado, Utah, USA;
Saskatchewan, Canada

🏛 Peabody Museum of
Natural History, New Haven, USA;
Academy of Natural Sciences,
Philadelphia, USA

Torosaurus, one of the most advanced of the long-shielded neoceratopsians, had a gigantic head that measured up to more than 8 feet (2.5 m) long. This means that its skull was longer than that of any other land animal that has ever lived. Torosaurus's neck frill, too, was enormous and made up about one-half the total skull length. The frill had two large, symmetrical openings that reduced its weight. Torosaurus had two prominent horns above its eyes and a very small horn on its snout. All three horns pointed forward.

In terms of length, Torosaurus was not much smaller than its contemporary, and close relative, Triceratops, but because of its more slender build, it probably weighed considerably less. The fact that relatively few specimens of Torosaurus have come to light suggests that this dinosaur was much less abundant than Triceratops.

John Bell Hatcher discovered a skull of Torosaurus in Niobrara County in Wyoming in 1889. He sent the specimen to Charles Othniel Marsh, who in 1891 named it Torosaurus latus, in recognition of the bull-like size of its skull and its large eye horns. This skull has been in the possession of the Peabody Museum in New Haven, Connecticut, ever since.

Partial remains of another four Torosaurus skulls have been found and have resulted in the naming of several species. However, differences between them have since been attributed to male–female variations.

Like other neoceratopsians, Torosaurus was a plant-eater that sheared off tough plant matter with a sharp beak powered by its strong shearing jaws. It then ground this food with the many rows of teeth in the back of its mouth.

An interesting study of a Torosaurus frill, undertaken in the 1930s by Dr. R. Moodie, showed irregular holes in the bony surface that may have been caused by cancers.

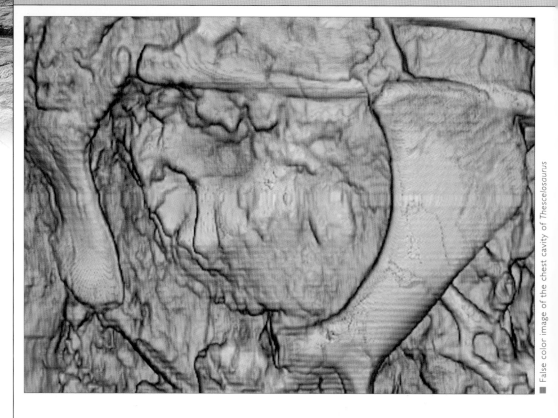

■ False color image of the chest cavity of *Thescelosaurus*

RECENT DINOSAUR DISCOVERIES

Our understanding of dinosaurs and their world is continually being expanded. Theories are constantly being re-evaluated as new finds come to light and ever more sophisticated technology is developed for examining and assessing both newly discovered and long-known material. In very recent years, a number of important discoveries have been made around the world. They include some spectacular complete skeletons and the first discovery of fossilized soft tissue for some dinosaurs.

In April 2000, the heart of a small ornithischian dinosaur, *Thescelosaurus*, was described. This was an exciting first for paleontology. The dinosaur, which was found in 1993 in South Dakota, was purchased by the North Carolina Museum of Natural Sciences in 1996. The heart was preserved within an ironstone nodule inside the animal's chest cavity. Its preservation was probably the result of a rare process, whereby iron-bound oxygen in the heart muscle mineralized while it was in contact with groundwaters. A CAT scan of

the heart showed that it was a four-chambered heart with a single systematic aorta. The overall structure of the heart suggests that it is inter-mediate in form between those of crocodiles and birds. It provides good evidence that some dinosaurs had high metabolic rates—in other words, that they were "warm-blooded."

Another remarkable recent example of soft tissue preservation was an almost complete small theropod, *Scipionyx*, that was found in the Italian Alps. The preserved remains of this dinosaur included its intestines and other parts of its soft anatomy.

The rich quarries at Liaoning in northern

Scipionyx

China continue to produce spectacular feathered dinosaurs. The feathered dromaeosaurid *Sinornithosaurus* was recently described, along with a feathered therizinosaurid, *Beipiaosaurus*, and two very birdlike dinosaurs, *Protarchaeopteryx* and *Caudipteryx*, both of which had feathers with an advanced branching structure similar to that of birds. These dinosaurs also showed skeletal links to the first birds. Several species of primitive birds have also been described from Liaoning.

Beipiaosaurus

The late Cretaceous of Montana has recently yielded a well-preserved complete skeleton of a very small dromaeosaurid, *Bambiraptor*. This dinosaur seems to have more features in common with birds than any other known dinosaur species and it had the largest brain of any dinosaur relative to its body weight. It was about 3 feet (1 m) long and probably weighed about 6 pounds (3 kg).

New kinds of ankylosaurs have recently been announced from the late Cretaceous of North America and China. *Nodocephalosaurus*, from New Mexico, is known from a relatively complete top of the skull 15 inches (39 cm) long. *Shanxia*, based on a partial skull, was found in Shanxi Province, northern China.

The giant tyrannosaurid from the Gobi Desert that was formerly called *Tarbosaurus bataar* has been assigned to a new genus, aptly called *Jenghizkhan* because it was so formidable. It differed from *Tarbosaurus* in that it was much larger, had more robust bones in the skull around the orbit, and relatively higher neck vertebrae. The name *Tarbosaurus* is now restricted to smaller forms—about 26 feet (8 m) long—of the species *T. efremovi*. *Byronosaurus* is another new dinosaur from the Gobi Desert. The skull of this small troödontid theropod was found by recent expeditions to Mongolia by New York's American Museum of Natural History and the dinosaur was described in 2000.

The abelisaurids, which are best typified by the horned *Carnotaurus* from Argentina, were an odd-looking group of theropods. In 1993, a team of scientists that included Catherine Forster, Scott Sampson, and David Krause found a perfectly preserved skull of another abelisaurid, *Majungatholus*, from the late Cretaceous of Madagascar. This dinosaur was described in 1997. Its skull resembled that of *Carnotaurus* in that the head was short and deep with short horns above the eyes and a median thick horn on top of the back of its head. These structures were probably used as ornamental displays for courtship.

Scott Sampson has also recently named two spectacular new neo-ceratopsians, or horned dinosaurs, *Einiosaurus* and *Achelousaurus*. Both of these were found from the late Cretaceous of Montana. *Einiosaurus* had a peculiar forward-bending nose horn and two frill horns that

In this holotype of the very birdlike dinosaur Protarchaeopteryx, the two arms, the tail with feathers, and the legs are clearly visible. The head is more difficult to make out. Some Chinese experts believe that Protarchaeopteryx was an ancestor of Archaeopteryx.

Byronosaurus

neck bones of a gigantic beast, *Superposeidon*, were described from the late Jurassic Morrison Formation in Utah. According to some estimates, it was substantially bigger than *Brachiosaurus*, a previous contender for the title of largest land animal. Some scientists believe that *Argentinosaurus* from South America may have been the largest animal to walk the Earth. Although it is known only from a few bones, *Argentinosaurus*'s huge thoracic vertebrae were up to 4 feet (1.2 m) high and this dinosaur may have weighed much more than 70 tons (71.5 t).

Gondwanatitan is a newly described sauropod, based on a partial skeleton from the late Cretaceous of southern Brazil. It was a small titanosaur—a group that had bony armored plates in their skin—and many have been a mere 20–23 feet (6–7 m) long. *Tehuelchosaurus* was discovered from Chubut Province in southern Argentina and was formally named in 1999. Known from about half the skeleton, its fossil preserved skin impressions that showed small hexagonal platelets of bone. It was about 50–60 feet (15–18 m) long.

Irritator, a new theropod recently described from Brazil, is thought to have been a spinosaurid—a member of the group that may have used their long snouts and hooked claws to catch fish. The spinosaurids seem to have been most abundant in the early Cretaceous of Gondwanan countries.

projected to the rear. *Achelousaurus*, too, had two rearward-pointing frill horns, but it had only a bony thickening on the snout and over the eyes—perhaps to support a thick outgrowth of roughened skin or keratinous horns.

In recent years Australia has been an important source of new dinosaur discoveries. *Ozraptor*, based on one distinctive tibia (shin bone) from the middle Jurassic of Western Australia, is the oldest theropod so far known from Australia. Its advanced astralagus (ankle bone) structure, as observed from impressions on the front of the tibia, has led some paleontologists to suggest that it could be the oldest member of the abelisaurid lineage. There are new theories about another Australian dinosaur—the enigmatic *Rapator* from the early Cretaceous of New South Wales. Studies of the single hand bone that has been found suggest that it may have been a giant alvarezsaurid, one of a group of fast-running birdlike dinosaurs that had a single large claw on each hand.

The largest of all land animals were the gargantuan long-necked sauropods. New finds suggest that we may not yet have found the biggest of them. In April 2000, the

Lirainosaurus astibiae (meaning "Astibia's slender lizard") is a newly described gracile titanosaurid from the late Cretaceous of Spain. It is known from vertebrae, a fragmentary skull, teeth, a shoulder bone, elements of the limbs and pelvis, as well as from armored scutes that seem to have been present in many Cretaceous sauropods.

Yet another new titanosaur is *Tangvayosaurus hoffeti* from the early Cretaceous deposits at Tang Vay, Savannakhet Province, in Laos. The species name honors Josue-Heilmann Hoffet, who had discovered dinosaurs in Laos in the 1940s and published brief accounts of them. The animal is known from two incomplete skeletons. *Tangvayosaurus* seems to have been a primitive titanosaur, 50 feet (15 m) long, that was related to the Thai sauropod *Phuwiangosaurus*.

Majungatholus

Huge fossil bone from *Argentinosaurus*

Superposeidon

Irritator

Many new finds have been made in Africa. Dr. Paul Sereno and his team described two new sauropods from the Niger Republic—*Jobaria tiguidensis* and *Nigersaurus taqueti*. These creatures lived in the lush early Cretaceous swamp forests that existed in what is now the Sahara Desert.

Jobaria is named after "Jobar," an animal from Tuareg mythology, while its species name refers to the fossil locality. It is one of the best understood of all sauropods: Several skeletons, of both adult and immature animals, have been found. It was huge—over 70 feet (21 m) long—but for a Cretaceous sauropod, it was surprisingly primitive. Its neck was quite short, while the teeth were broad and chisel-like, rather like those of *Camarasaurus* from the late Jurassic of North America.

Nigersaurus taqueti (meaning "Taquet's lizard from Niger") was a smaller but more specialized animal than *Jobaria*. It was a 50-foot (15 m) long member of the rebbachisaurids—the group thought to be late surviving relatives of *Diplodocus*. *Nigersaurus*'s many teeth were small, with enamel on only one side, and were arranged in a tight battery at the front of the face. The snout was broad and squared off.

Atlasaurus imelakei was named from the Arabic word for giant (*imelake*) and its site, the Altas Mountains. It is based on an almost complete skeleton from the middle Jurassic of Morocco. This sauropod was about 45 feet (14 m) long. It was probably related to *Brachiosaurus*, but it had a relatively shorter neck and more elongated legs and tail.

CHAPTER SIX

A GUIDE TO DINOSAUR SITES AND MUSEUMS

USING THE GUIDE TO DINOSAUR SITES AND MUSEUMS

Dinosaur fossils are often found in remote locations that few can visit.

Fortunately there are numerous museum collections open to public view.

The following pages feature a selection of the world's most famous dinosaur fossil sites as well as museums that house and display important collections of dinosaur fossils. Many of the museums promote the science of paleontology and are a good source for finding out more about the history of dinosaur finds and ongoing research programs.

The illustrated banding *at the top of the page is a visual pointer to indicate that the page is about a site or a museum.*

Dinosaur Sites and Museums

Color plates *show a detail of the landscape of a site and a particular find related to that site, or the façade of a museum, and an interior display area.*

Field Notes Panels
* Location of the site
* Period to which a dinosaur fossil site belongs
* Names of the dinosaurs discovered at the site or on display in the museum
* Names of the main museums that display dinosaur fossils from the site featured

The text *describes the features of a site and how it was discovered. In the case of the museum, it details the history of the institution, its importance to the study of paleontology, and the key people involved in its development.*

Dinosaur National Monument
Utah, United States of America

The United States National Parks Service established Dinosaur National Monument on October 4, 1915. Its establishment was the result of a decree by President Woodrow Wilson, which was designed to protect this extraordinary late Jurassic site from the depredations of unscrupulous fossil-hunters. This decree was prompted by an unsuccessful attempt on the part of the Carnegie Museum of Natural History in Pittsburgh, Pennsylvania, to file a claim to the site, which is located just north of the Jensen, and close to the border between Utah and Colorado.

The site was discovered in 1909 when Earl Douglas of the Carnegie Museum noticed the skeleton of a large sauropod eroding out of an exposed sandstone ledge on the banks of the Green River. Scientists from the Carnegie Museum soon began excavation work, and a seemingly endless succession of fossils was brought to light. So many bones were found at the site that field work continued without interruption until 1923. The final outcome of these excavations was that approximately 350 tons (357 t) of fossil bones were shipped back to the Carnegie Museum for preparation and study. The specimens contained the most complete skeleton of the sauropod *Apatosaurus* ever found (this was the very specimen that was 214 found by Douglas in 1909). Charles Whitney

The entrance to Utah's Dinosaur National Monument.

Gilmore, of the Smithsonian Institution, worked on this skeleton for many years, finally publishing a detailed monograph describing *Apatosaurus* in 1936.

The lengthy excavations also uncovered nearly complete skeletons of *Allosaurus, Camptosaurus, Dryosaurus,* and *Stegosaurus,* all of which eventually became mounted displays at the Carnegie Museum. After the museum's expeditions had finished at the site, work continued over the next two decades, first by parties from the Smithsonian, under the auspices of Gilmore, and then by teams from the University of Utah. The Dinosaur National Monument we see today is an exposed wall of sandstone that has

208

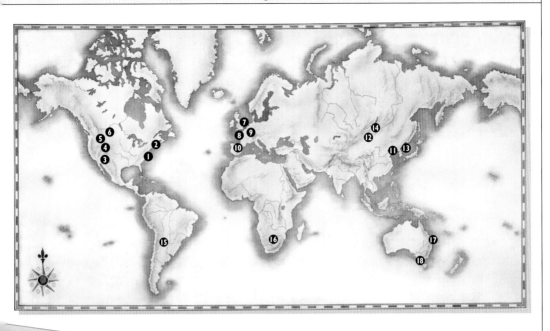

Dinosaur National Monument

SITES AND MUSEUMS

1. Smithsonian Institution, Washington DC, USA
2. American Museum of Natural History, New York, USA
3. Dinosaur National Monument, Utah, USA
4. Hell Creek, Montana, USA
5. Dinosaur Trackways of the Western United States, USA
6. Dinosaur Provincial Park, Alberta, Canada
7. Natural History Museum, London, UK
8. Musée National d'Histoire Naturelle, Paris, France
9. Holzmaden and Solnhofen, Germany
10. Las Hoyas, Spain
11. Institute of Vertebrate Paleontology and Paleoanthropology, Beijing, China
12. Zigong Dinosaur Park, Szechuan, China
13. Liaoning, China
14. Flaming Cliffs, Mongolia
15. Valley of the Moon, Argentina
16. Karoo Basin, South Africa
17. Lark Quarry, Queensland, Australia
18. Dinosaur Cove and East Gippsland, Victoria, Australia

—efully excavated to reveal
—00 dinosaur bones. They
—n prepared in high relief
—cur naturally within the
— of the Morrison
—. This wall is within the
—itor Center. Opened in
—enter completely
— wall and protects it.
—ows a partial skeleton
—*Camarasaurus*, along
— bones of several
—rs, such as *Apato-*
—*rus*, *Allosaurus*, and *Dryosaurus*.
— well preserved for their age,
—een 155 and 148 million years.
—different genera of dinosaurs have
—n the site. This means that it
—st diverse late Jurassic dinosaur
—e in the world.
—Formation is exposed through-
—th length of the United States.
—rops indicate a dry desert en-
—s around Utah the sequence
—nosaurs there once inhabited
—rge, lowland alluvial plain.

—us stands in front of Dinosaur
—ry Visitor Center.

FIELD NOTES
South-western United States
Late Jurassic
Allosaurus, large sauropods, Stegosaurus, Camptosaurus
University of Utah, USA; Cleveland Museum of Natural History, USA; replica skeletons (e.g. Diplodocus, Allosaurus) in many of the world's major museums
Open to visitors as a national park

The richness of the fossil finds suggests that the dinosaurs were probably swept away and drowned by large, meandering rivers during floods. At the bends where the river currents slowed down, the carcasses would have been dumped, gradually forming a mass burial site. Later on, the bones of the decaying carcasses would have been scattered by scavenging predators or moved around by the action of the water, resulting in the random jumble of different dinosaur remains, some partially articulated, that are exposed for visitors to the site to see today.

Tyrannosaurus exhibit

Smithsonian Institution

Washington DC, United States of America

The Smithsonian Institution was established in 1846 from a private bequest by James Smithson. Today its National Museum of Natural History contains almost 50 million fossil specimens, held in 450 separate collections. Its dinosaur collection, started at the beginning of the 20th century with specimens assembled under the auspices of Charles Othniel Marsh, includes more than 1,500 specimens. There are examples of late Jurassic dinosaurs such as *Allosaurus*, *Stegosaurus*, *Ceratosaurus*, and *Diplodocus*, collected from the Morrison Formation in Wyoming and from various sites in Colorado. Also, the collection features many skulls of *Triceratops* found by the famous dinosaur hunter John Bell Hatcher from the late Cretaceous deposits of Lance Creek, Wyoming.

Under the direction of Charles Whitney Gilmore, late Cretaceous dinosaurs were systematically collected from New Mexico and Montana. Gilmore's early finds from the Two

Medicines Formation in northern Montana included the horned dinosaur *Brachyceratops* as well as bone beds that contained juvenile hadrosaurids and many other kinds of vertebrates.

Other dinosaurs collected under Charles Whitney Gilmore include the only titanosaurid sauropod that is known from North America (*Alamosaurus*, from New Mexico) and a range of interesting specimens that reveal pathological abnormalities in dinosaur bones.

Gilmore was one of the most productive of the early 20th-century dinosaur scientists. His work describing the nearly complete skeleton of *Apatosaurus*, first located by Earl Douglas in 1909, was critical to the forming of the Dinosaur National Monument in Utah. The Smithsonian Institution worked the sites in this area during the 1920s and 1930s, extracting many fine dinosaur specimens.

This reconstruction of Stegosaurus is a popular exhibit in the Smithsonian Institution.

Reconstruction of Antrodemus

By 1940 the Institution had become home to one of the largest collection of dinosaurs in the world. Gilmore wrote a summary paper—published in 1941 in the museum's proceedings—detailing the history and development of the Smithsonian's remarkable vertebrate paleontology collection.

The Smithsonian's collection of dinosaur type specimens contains over 40 named species. Many of the earlier ones, though, have since been dropped as valid species names or recognized as belonging to other known species. *Dystrophaeus viaemalae*, the first type dinosaur of the Institution was collected by John Strong Newberry in 1859 and named by Edward Drinker Cope in 1877. However, these fossils are now thought to belong to *Camarasaurus*, whose name takes priority.

Some of the museum's most important specimens are the skeletons of *Alamosaurus sanjuanensis*, *Stegosaurus stenops*, *Brachyceratops montanus*, *Edmontia rugosidens*, *Pachycephalosaurus wyomingensis*, *Camptosaurus brownei*, *Ceratosaurus nasicornis*, *Edmontosaurus annectens*, *Styracosaurus ovatus*, and *Thescelosaurus neglectus*.

The permanent exhibitions of the Smithsonian's Natural History Museum include several mounted dinosaur skeletons containing more then 50 percent original material. Among these exhibits are *Albertosaurus*, *Allosaurus*, *Camarasaurus*, *Diplodocus*, *Ceratosaurus*, *Corythosaurus*, *Edmontosaurus*, *Stegosaurus*, and *Triceratops*.

As well, there are mounted casts of the skeletons of a baby *Maiasaura* and the small South African dinosaur *Heterodontosaurus*. Other interesting material on display includes the skull of *Centrosaurus* and a cast of a *Tyrannosaurus* skull from the collections of the American Museum of Natural History.

Today, the Smithsonian's Department of Paleobiology is a major research body for all aspects of paleontology—marine microfossils, marine mollusks, fossil whales, and plant evolution, as well as dinosaurs. Further detailed information about the work of the staff, the collections, and the research programs being held can be found on the Smithsonian's web site.

A lifelike reconstruction in the Smithsonian of the giant pterosaur Quetzalcoatlus with its wings spread in flight.

FIELD NOTES

🐾 *Superb displays of late Cretaceous theropods, hadrosaurs, and neoceratopsians, and late Jurassic dinosaurs from the Morrison Formation in Wyoming*

211

The AMNH's main entrance

American Museum of Natural History

New York, United States of America

The American Museum of Natural History (AMNH), in New York City, was first incorporated in 1869. Now the largest private museum in the world, AMNH houses a huge collection of dinosaur specimens, including the largest collection of real dinosaur material anywhere in the world. More importantly, it is a major research center for work on dinosaur systematics and evolution.

The AMNH began conducting field trips in 1897 with the early expeditions concentrating on the rich dinosaur localities found around Como Bluff, Wyoming. The team of eminent paleontologists included Henry Fairfield Osborn, Barnum Brown, William Matthew, Walter Granger, Jacob Wortman, and Albert Thomson. They made important finds of the sauropods *Diplodocus* and *Apatosaurus* (which was formerly known as *Brontosaurus*), and of the theropods *Allosaurus* and *Ornitholestes*. In 1898, the AMNH expeditions discovered the famous Bone Cabin site in Wyoming, where more than 50 partial dinosaur skeletons were collected. They included the remains of *Camarasaurus*, *Camptosaurus*,

> **FIELD NOTES**
>
> 🦖 *A wide range of dinosaur exhibits, both saurischian and ornithischian. Spectacular skeletons include* Tyrannosaurus, Diplodocus, Triceratops, *and* Euoplocephalus

Dryosaurus, and *Stegosaurus*. As well as these, they found new fossils of the species that they had unearthed earlier at Como Bluff.

In 1902, Barnum Brown, an eccentric character who often wore elaborate fur coats in the field, led an AMNH expedition to the Hell Creek region of Montana where he discovered the first *Tyrannosaurus rex* specimen. Later at this site—in 1908—he found an almost complete *Tyrannosaurus*, the mounted skeleton of which has been on display at the AMNH for more than 50 years. Between 1910 and 1915 Brown scoured the Red Deer River region of Alberta, Canada, using a custom-built wide barge that could cruise on the river and allow him access to remote sites. From these expeditions, he brought back substantially complete skeletons of the hadrosaurs *Corythosaurus* and *Saurolophus*. In the 1930s, he returned to the Jurassic deposits of Wyoming and collected many other dinosaur specimens for the AMNH.

Between 1922 and 1925, AMNH teams, under the leadership of the adventurer–scientist Roy Chapman Andrews and including Walter

212

Tyrannosaurus skeleton

Jacob Wortman, Walter Granger, and Peter Kaiser unearthing Apatosaurus *from the Bone Cabin Quarry site in Wyoming.*

Granger and Henry Osborn, ventured much farther afield to explore the rugged Gobi Desert of Mongolia. These now famous expeditions were originally undertaken in order to uncover fossil evidence of early humans; instead, they brought back a fine collection of late Cretaceous dinosaurs, including the first discovery of a complete nest of dinosaur eggs—belonging to *Protoceratops*. The Flaming Cliffs region was found to be particularly productive. Here, the AMNH field parties collected specimens of *Velociraptor*, *Protoceratops*, *Oviraptor*, *Pinacosaurus*, and *Saurornithoides*.

During the 1990s, teams from the AMNH made regular visits to these sites in the Gobi Desert, where they worked with paleontologists from the Mongolian Academy of Sciences. These recent expeditions have resulted in many new

and extraordinary dinosaur, bird, and mammal discoveries, including specimens of the theropod *Oviraptor* brooding over a nest of its eggs. Remarkably, some of these eggs have yielded well-preserved embryonic dinosaur skeletons.

The AMNH's new Halls of Saursichian and Ornithischian Dinosaurs, which were completed in the mid-1990s, display more than 100 dinosaur specimens. About 85 percent of these—including the famous skeletons of *Tyrannosaurus*, *Apatosaurus*, *Triceratops*, and *Euoplocephalus*—consist entirely of original material. The collections of the AMNH form the basis of its highly successful research programs and include many thousands of important dinosaur fossils, large quantities of which were collected years ago and are still awaiting preparation and classification.

In recent years, the AMNH has produced a series of guide books that illustrate and give details of all the important specimens on display. It has also produced a number of large-format illustrated books about the history and development of the museum's vast collections.

213

Theropod footprints within the Morrison Formation in Utah

Dinosaur National Monument

Utah, United States of America

The United States National Parks Service established Dinosaur National Monument on October 4, 1915. Its establishment was the result of a decree by President Woodrow Wilson, which was designed to protect this extraordinary late Jurassic site from the depredations of unscrupulous fossil-hunters. This decree was prompted by an unsuccessful attempt on the part of the Carnegie Museum of Natural History in Pittsburgh, Pennsylvania, to file a claim to the site, which is located just north of the Jensen, and close to the border between Utah and Colorado.

The site was discovered in 1909 when Earl Douglas of the Carnegie Museum noticed the skeleton of a large sauropod eroding out of an exposed sandstone ledge on the banks of the Green River. Scientists from the Carnegie Museum soon began excavation work, and a seemingly endless succession of fossils was brought to light. So many bones were found at the site that field work continued without interruption until 1923. The final outcome of these excavations was that approximately 350 tons (357 t) of fossil bones were shipped back to the Carnegie Museum for preparation and study. The specimens contained the most complete skeleton of the sauropod *Apatosaurus* ever found (this was the very specimen that was found by Douglas in 1909). Charles Whitney

The entrance to Utah's Dinosaur National Monument.

Gilmore, of the Smithsonian Institution, worked on this skeleton for many years, finally publishing a detailed monograph describing *Apatosaurus* in 1936.

The lengthy excavations also uncovered nearly complete skeletons of *Allosaurus*, *Camptosaurus*, *Dryosaurus*, and *Stegosaurus*, all of which eventually became mounted displays at the Carnegie Museum. After the museum's expeditions had finished at the site, work continued over the next two decades, first by parties from the Smithsonian, under the auspices of Gilmore, and then by teams from the University of Utah.

The Dinosaur National Monument we see today is an exposed wall of sandstone that has

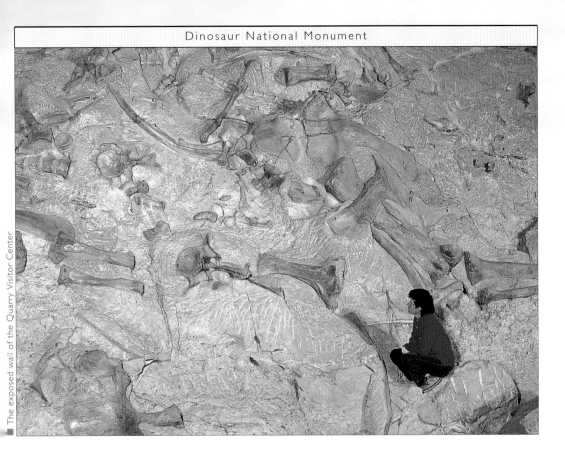

The exposed wall of the Quarry Visitor Center

been carefully excavated to reveal some 1,500 dinosaur bones. They have been prepared in high relief as they occur naturally within the sandstones of the Morrison Formation. This wall is within the Quarry Visitor Center. Opened in 1958, the center completely encloses the wall and protects it. The wall shows a partial skeleton of a juvenile *Camarasaurus*, along with isolated bones of several other dinosaurs, such as *Apato-saurus*, *Diplodocus*, *Allosaurus*, and *Dryosaurus*. The bones are well preserved for their age, estimated between 155 and 148 million years.

To date, 10 different genera of dinosaurs have been found from the site. This means that it stands as the most diverse late Jurassic dinosaur locality anywhere in the world.

The Morrison Formation is exposed through-out the north–south length of the United States. In the south, outcrops indicate a dry desert en-vironment, whereas around Utah the sequence suggests that the dinosaurs there once inhabited an area that was a large, lowland alluvial plain.

FIELD NOTES

South-western United States

Late Jurassic

Allosaurus, large sauropods, Stegosaurus, Camptosaurus

University of Utah, USA; Cleveland Museum of Natural History, USA; replica skeletons (e.g. Diplodocus, Allosaurus) in many of the world's major museums

Open to visitors as a national park

The richness of the fossil finds suggests that the dinosaurs were probably swept away and drowned by large, meandering rivers during floods. At the bends where the river currents slowed down, the carcasses would have been dumped, gradually forming a mass burial site. Later on, the bones of the decaying carcasses would have been scattered by scavenging predators or moved around by the action of the water, resulting in the random jumble of different dinosaur remains, some par-tially articulated, that are exposed for visitors to the site to see today.

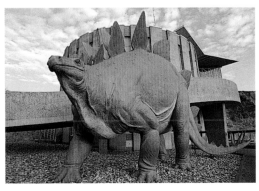

A reconstructed Stegosaurus *stands in front of Dinosaur National Monument's Quarry Visitor Center.*

Barnum Brown surveys a fossil at Hell Creek

Hell Creek

Montana, United States of America

The exposures at Hell Creek represent a sequence of late Cretaceous river deposits that have been dated to between 70 and 65 million years ago (the late Maastrichtian stage)—the very last years of the age of dinosaurs. In 1902 Barnum Brown from the American Museum of Natural History (AMNH) first began searching the Hell Creek area for dinosaurs. He named the sedimentary rocks exposed there as the Hell Creek Beds, and today the 450-foot (137 m) thick succession of sandstones, siltstones, and mudstones are known formally as the Hell Creek Formation. These outcrops are protected within Hell Creek State Park, situated north of Jordan, Montana. In 1902, Brown discovered the first, and largely incomplete, *Tyrannosaurus* skeleton at Hell Creek and in 1908 he found an almost complete one, as well as the first skull of *Triceratops* from the region.

Part of the Hell Creek Formation in Montana.

In the years since then, the Hell Creek Formation has been searched by many field parties and has produced a great number of splendid dinosaur fossils, including, among others, *Torosaurus, Edmontosaurus, Stygimoloch, Pachycephalosaurus, Stegoceras, Albertosaurus, Aublysodon, Ornithomimus, Troödon,* and *Ankylosaurus*. The pachycephalosaur *Stygimoloch* was actually named for the Hell Creek site, its name

A collector at work in the Hell Creek Formation, where many species of fossil plants and animals have been found.

The skeleton of Thescelosaurus, found in the Hell Creek Formation

The distinctive domed skull of Stygimoloch.

meaning literally "river of Hell devil." Although many species of dinosaur are present here, very few are found in any degree of articulation. Most of the dinosaurs of Hell Creek are smaller animals, which suggests the area may have been fairly densely forested. As well as dinosaurs, the Hell Creek Formation has yielded the remains of several different kinds of tortoises, giant monitor lizards, and tree-dwelling multituberculate mammals.

The Hell Creek region is also of scientific significance because the upper few yards of its sedimentary rocks possibly date into the Paleocene (the earliest stage of the Cenozoic era). High concentrations of the rare trace element iridium (an element found in abundance in meteorites) have been found in a layer here. This has been suggested as evidence for a giant meteorite impact at the end of the Cretaceous, an event that may have been the main cause of the extinction of the dinosaurs. The controversial occurrences of dinosaur remains within Paleocene deposits at Hell Creek has led some paleontologists to suggest that dinosaurs could have survived the meteorite

FIELD NOTES

North-western United States

■ Late Cretaceous

Triceratops, Tyrannosaurus, hadrosaurs, pachycephalosaurs

Denver Museum, Colorado, USA; Carnegie Museum, Pittsburgh, USA; Smithsonian Institution, Washington DC, USA; American Museum of Natural History, New York, USA

■ Hell Creek State Park is.

impact and lived on for a short while in the early part of the Paleocene. Others, however, interpret these occurrences as being nothing more than the re-siting of Cretaceous remains into Paleocene layers as a consequence of the erosion of the Cretaceous deposits.

The Hell Creek exposures have produced more than 190 species of fossil plants, and the microflora (based on pollen and spore species) suggests that up to 300 different species of plants grew in the floodplain and surrounding mountain settings. In the dry, subtropical conditions that prevailed there, the landscape would have been dominated by flowering plants.

Many of the Hell Creek dinosaurs went into the collection of the Denver Museum of Natural History, and displays there feature skeletons of *Tyrannosaurus* and *Edmontosaurus*, the latter showing fossilized evidence of a wound that may have been inflicted by the former.

217

Jurassic sauropod trackways in the Morrison Formation, Colorado

Dinosaur Trackways of the Western United States

United States of America

Dinosaur trackways are relatively common fossils and are found in nearly all parts of the world. They date back to most parts of the Mesozoic era. Dinosaur footprints were first described as long ago as 1858 by the English clergyman and paleontologist the Reverend Edward Hitchcock. Hitchcock allocated names to the different kinds of dinosaur tracks he identified, and some of these, such as the large Triassic theropod tracks called *Eubrontes*, are still widely used today.

The best preserved dinosaur footprints are those that were formed when animals walked along a flat, wet, sandy area, such as a coastal plain or large river mouth delta. The feet left impressions in the wet sandstone which baked hard in the sun, later becoming buried by fine-grained, wind-blown, or water-borne sediments. The layer containing the footprints would have turned to hard, sedimentary rock once it had been buried by the deposition of minerals between the sediment grains. At a later time, the uplift of the Earth's crust and erosion would have once again exposed the footprint-bearing layer at the surface.

Trackways can tell us a great deal about dinosaur behavior—such as how the animals walked or ran, at what speeds they traveled, and the nature of dinosaur herd behavior. In some instances, they provide insights into the nature of dinosaur predatory habits.

The western United States has a large number of significant trackway sites, ranging from the late Triassic through to the late Cretaceous. Most of these occur in river, lake, shoreline, and desert dunefield settings.

Triassic dinosaur trackways occur in the Chinle Formation of New Mexico, Arizona, and Colorado. They comprise mainly the footprints of small coelurosaurs such as *Coelophysis*, and were left in riverside sediments. Late Triassic or early Jurassic dinosaur trackways occur in southern Utah from the Warner Valley south of Zion National Park and near Tuba

A large early Jurassic theropod footprint in the Kayenta Formation, near Tuba City, Arizona.

Turning sauropod trackways in the Morrison Formation, Utah

City, Arizona. In the Moenave and Kayenta formations in this region, there are well-preserved, three-toed prints of theropods that have been given the names *Grallator, Kayentopus,* and *Eubrontes. Grallator* tracks may have represented a small dinosaur similar to *Coelophysis; Eubrontes* may have been a large plateosaurid prosauropod; *Kayentopus* appears to have been a medium-sized theropod—it may have been *Dilophosaurus*, fossils of which occur in the same rock layers. The top layers of the Entrada Sandstone, which outcrops prominently in the Arches National Park, northern Arizona, is rich in dinosaur trackways. The late Jurassic Morrison Formation has some layers that contain many trackways. Perhaps the most celebrated is the Purgatoire Valley, south-eastern Colorado, where the tracks are preserved in rocks that were once an ancient shoreline. The trackways representing some 1,300 prints, from around 100 dinosaurs—principally theropods and sauropods, but including some rarer ornithopods—are recorded.

FIELD NOTES

☀ Central western United States; western Canada

■ Late Triassic to late Cretaceous

🦋 Many kinds of sauropods, theropods, ornithopods, and thyreophoran trackways

🏛 On-site trackways at Dinosaur Valley State Park, Texas, USA and Picketwire Canyonlands, Colorado, USA; Texas Memorial Museum, Austin, USA; American Museum of Natural History, New York, USA; Royal Tyrrell Museum, Alberta, Canada

■ Many sites open to visitors; others on private land require permission

The early Cretaceous Dakota Group of Colorado shows layers with a high density of footprints, which results in a trampled effect on the sedimentary layers. In the Laramie Formation of the Denver region, one set of layers of "trampled" rock made by ornithopod dinosaurs can be traced for tens of miles. Tracks of neoceratopsians and medium to small theropods are also recorded here.

The early Cretaceous Paluxy River trackways are exposed near Glen Rose in Texas (Glen Rose Formation). Large theropod tracks here were possibly made by *Acrocanthosaurus*; the largest sauropod tracks could have been made by *Pleurocoelus*. These trackways are famous for the insights they give into sauropod herd behavior. They show that the largest sauropods walked on the outside of the herd with the smaller juveniles positioned safely in the center of the group. Other trackways preserved here show possible attack trails made by large, predatory theropods closing in on sauropods.

219

Dinosaur Hall in the Royal Tyrrell Museum

Dinosaur Provincial Park

Alberta, Canada

The extensive outcrops of the late Cretaceous Judith River Group—which dates back to between about 77 and 73 million years ago—along the Red Deer River near central southern Alberta, have yielded a greater number of complete dinosaur specimens than any other site on Earth. In 1955, an area of 28 square miles (73 km²) of this land was officially established as Dinosaur Provincial Park, and in 1979 the region was designated as a World Heritage Site by UNESCO for its exceptional preservation, diversity, and the abundance of dinosaurs and related fauna. The deep gullies and river outcrops of the park were formed by erosion caused from glacial meltwaters retreating from the last ice age—between about 14,000 and 12,000 years ago—exposing a section 400 feet (125 m) thick of late Cretaceous sandstones, mudstones, and siltstones that are collectively called the Dinosaur Park Formation.

The Dinosaur Park Formation represents an estuary-river setting. Large numbers of dinosaur skeletons may have got caught up here when floating carcasses became entangled in accumulations of logs (log jams) at river bends. Massive

FIELD NOTES

South-western Canada

Late Cretaceous

Many dinosaur types, including ceratopsian and hadrosaur species; Albertosaurus, Tyrannosaurus

Royal Tyrrell Museum, Drumheller (on-site), and many museums worldwide

Open to visitors as a national park

flood events may have killed thousands of dinosaurs at one time, as can be seen in the *Centrosaurus* bone bed.

The history of Dinosaur Provincial Park dates back to 1909, when a rancher, John Wagner, reported to New York's American Museum of Natural History (AMNH) that he thought he had dinosaur bones on his property. In the next year or so, Barnum Brown started to search the area. Brown soon realized that the dinosaur-bearing rocks were best exposed along the banks of the Red Deer River. He therefore built a wide raft with a tent over the top of it so that his field crew could sail down the river, stopping to search at each new exposure. At the end of each field season, as the cold weather closed in on them, the fossil-hunters hauled the raft onto the banks and stored their finds until the next season.

By 1912 the boundaries of the park were established around the exposures. Brown's early finds included complete skeletons of *Gorgosaurus*, *Corythosaurus*, *Prosaurolophus*, *Centrosaurus*, and *Struthiomimus*. Collector Charles Sternberg and his family also worked the sites within the park.

■ An eroded rock formation in Dinosaur Provincial Park

The Royal Tyrrell Museum (left), one of the world's most important paleontological museums, is situated within Dinosaur Provincial Park.. The skull (above) of the horned dinosaur Pachyrhinosaurus is part of the museum's collection.

They found skeletons of *Albertosaurus* and *Chasmosaurus* with delicate skin impressions, as well as beautifully preserved skulls of *Centrosaurus* and *Styracosaurus*.

Field work has continued in the park over the succeeding years, and has yielded as many as 250 articulated dinosaur skeletons that represent some 36 different dinosaur species, including *Struthiomimus, Anchiceratops, Lambeosaurus, Dromaeosaurus, Troödon, Gravitholus, Cryposaurus, Brachylophosaurus, Edmontia, Euoplocephalus,* and *Panoplosaurus*. Approximately 300 different fossil species of animals and plants have been recorded for the Dinosaur Park Formation. This includes 84 species of fishes, frogs, salamanders, turtles, lizards, crocodiles, pterosaurs, birds, and mammals, as well as many species of fossil plants.

The Royal Tyrrell Museum of Paleontology established a permanent field station within the park in 1987. This provides a base for ongoing field studies for the collection of new dinosaur remains, and also functions as a center where visitors to the park can view displays and find information about the discoveries that were made there. The exhibits include the skeletons of the horned *Centrosaurus* and the large predator *Daspletosaurus*. Today, the dinosaurs that have been collected from the park can be viewed and studied in more than 30 institutions all around the world.

221

The Natural History Museum's impressive façade

Natural History Museum

London, United Kingdom

For the past two centuries London's Natural History Museum has housed the first dinosaur fossils ever found and described. As well, it has built up an enormous collection of fossils from around the world. The British Museum first opened in 1753 in the part of London known as Bloomsbury. There, it displayed a wide range of natural and historical curiosities. The museum's first significant collections of fossils and geological specimens were made up of both purchases and donations from the private collections of a number of eminent "gentlemen" naturalists of the time. Among the donors were William Smith—often referred to as the "father of geology"—and Gideon Mantell, who discovered remains of *Iguanodon*. All the specimens that Charles Darwin collected on his voyage on the *Beagle* came to reside in the museum.

The museum moved location in 1845, but by 1856 the collections were so large that another building was required. Specimens from the far corners of the British Empire came rolling in; among them were vast numbers of foreign fossils collected by the British Geological Survey. The current impressive building that stands in South

FIELD NOTES

🖤 *A huge range of important specimens from all around the world.* Prominent among them are Euoplocephalus, Diplodocus, Triceratops, Iguanodon, Baryonyx, Hypsilophodon, Brachiosaurus, Archaeopteryx, *and* Scelidosaurus.

Kensington was completed in 1881. One of the activists behind the push for the new museum, and a contributor toward its design, was the famous anatomist Sir Richard Owen. It was Owen who in 1842 coined the term "Dinosauria," after he undertook a thorough revision of all the British dinosaur remains that were known at the time. Owen's revision was based principally on far from complete specimens of *Megalosaurus*, *Hylaeosaurus*, and *Iguanodon*.

In 1963, the Natural History Section of the museum separated from the British Museum (whose collections are now predominantly historical and archaeological). It became known officially as the Natural History Museum. In 1986, its collections were increased when those belonging to the former Geological Museum, situated nearby, were incorporated into them.

Today the Natural History Museum holds some 9 million specimens, including approximately 30,000 specimens of fossil reptiles and amphibians. The museum is renowned as a center for research in all aspects of paleontology, including the study of dinosaurs. Among its most prominent dinosaur specialists over the past three

Mamenchisaurus skeleton

decades have been Dr. Alan Charig and Dr. Angela Milner.

The dinosaur collections of the Natural History Museum contain many important specimens such as *Megalosaurus, Baryonyx, Cetiosauriscus, Hylaeo-saurus, Hypsilophodon, Dacentrurus, Rhabdodon (=Mochlodon), Polacanthus, Euoplocephalus, Thecodontosaurus,* and *Brachiosaurus.* The London specimen of *Archaeopteryx*, perhaps one of the most famous fossils of its time, and one of only seven known specimens, was acquired by the museum in 1861 for what was then considered a princely sum— £700 ($1,070). Sir Richard Owen's study of this specimen was a seminal work on the origin of birds, even though nowadays *Archaeopteryx* is generally regarded as the first bird by definition, even though it is more dinosaurian by design.

Another of the museum's most famous dino-saurs is the small, armored *Scelidosaurus* which Sir Richard Owen described from a nearly com-plete skeleton found in Dorset, in the south of England, in 1863. In the 1960s, technicians at the Natural History Museum developed the very valuable technique of preparing fossil bones out of limestone rock by using weak solutions of acid. By this means, *Scelidosaurus* has now been

This specimen is a jumble of Iguanodon *bones known as the "Maidstone Slate." It was developed in 1834 by the eminent naturalist Gideon Mantell..*

carefully prepared to more clearly display many details of its anatomy.

The Natural History Museum also contains type specimens of many very poorly known dinosaurs, such as the opalized bones of *Rapator, Walgettosuchus,* and *Fulgurotherium* from Australia. In addition there are numerous other named dinosaurs in the collections whose validity is under question, such as the type specimen of the theropod *Aristosuchus,* from the Wealden of England, and *Proceratosaurus* from the Jurassic of southern England.

The dinosaur display at the Natural History Museum was updated in 1991 in order to show dinosaurs in the light of their reconstructed biology and behavior. Visitors walking into the main entrance hall of the museum find them-selves in the presence of impressive mounted skeletons of *Diplodocus* and *Triceratops.*

223

La Galérie de Palinto

Musée National d'Histoire Naturelle

Paris, France

The Institut de Paléontologie, part of France's Musée National d'Histoire Naturelle, is situated in the beautiful setting of the Jardin des Plantes in Paris. The museum houses the large comparative anatomy collection of the great French anatomist Baron Georges Cuvier (1769–1832) together with a great number of important fossil specimens from around the world. It has many dinosaur specimens, especially from France, Africa, and parts of South-East Asia.

Triassic dinosaurs in the institute include French material of the prosauropod *Plateosaurus* (from Saint Lothan), an undescribed prosauro-pod from Alzon, and Moroccan specimens of *Azendohsaurus*. From 1955 to 1959, expeditions to Lesotho, in southern Africa, under the leadership of Dr. L. Ginsberg and Drs. F. and P. Ellenberger, collected specimens of the large prosauropod *Euskelosaurus* and a lower jaw of the ornithischian *Fabrosaurus* from Maphutseng.

Jurassic dinosaurs in the institute include several local species. The theropod *Poekilo-pleuron bucklandii* is now known only from casts, as the original bones, first described in 1838,

were destroyed in World War II. The theropod *Piveteausaurus* is known from a braincase found at Calvados in Normandy and a partial sauropod skeleton, *Bothriospondylus*, came from Damparis.

Baron Cuvier described the bones of a reptile from Honfleur, which he thought was a large crocodile but which was later identified as a theropod dinosaur and named as *Streptospondylus cuvieri*. The most complete Jurassic dinosaur known from France is a small skeleton of the little theropod *Compsognathus corallestris* from the area around Canjuers.

The institute contains some large Jurassic sauropod bones from Africa and Madagascar. In 1940–1941, Dr. A. de Lapparent collected *Cetiosaurus* from El Mers, Morocco, and in 1962, Dr. L. Ginsberg collected *Lapparentosaurus* and *Bothriospondylus* from Kamoro, Madagascar.

Cretaceous dinosaurs are generally poorly known, some by only a few, isolated bones. The institute's collections include the theropod *Genusaurus* from Bevons, a hadrosaur jaw bone from Saint Martory, and from the latest Cretaceous deposits of Fox Amphoux, specimens that were collected in 1939 by

FIELD NOTES

🌿 Iguanodon, Allosaurus, Diplodocus, Tyrannosaurus, Tarbosaurus

224

The museum façade

Dr. de Lapparent. These include part of a nodosaurid as
well as titanosaur sauropod
bones, hypsilophodont bones,
and some theropod teeth and
bones. Unfortunately, the start
of World War II put a stop to
de Lapparent's productive
dinosaur excavations at this site.
Dinosaur eggs from the late
Cretaceous were also collected
from Bouches de Rhône, in southern France.

The institute has a good representation of
Cretaceous dinosaurs from Africa. In 1950,
Dr. Lavocat first collected bones and teeth of the
giant theropod *Carcharodontosaurus* from Kem-
Kem in Morocco. In 1995, an almost complete
Carcharodontosaurus skull was found by Dr. Paul
Sereno's team from the University of Chicago.
Lavocat also found bones of the sauropod
Rebbachisaurus from Morocco, and Dr. Philippe
Taquet led major field expeditions into the
Niger Republic, Africa, between 1966 and 1972
and discovered the nearly complete skeleton of
the hump-backed iguanodontid *Ouranosaurus*.
The institute holds only a cast of this dinosaur;
the original specimen resides in the National
Museum of Niger. Other finds from these
expeditions held in Paris include a femur of the

*Professor Philippe Taquet is a former
director of the museum.*

dryosaurid *Valdosaurus* and
bones of the spined theropod
Spinosaurus.

The first dinosaurs collected
from the rich Cretaceous sites
of Madagascar are part of the
institute's collection. These
include bones of the sauropod
Titanosaurus madagascarensis, and a peculiar piece
of a domed skull that was originally identified
as belonging to a pachycephalosaur. In 1998, a
complete skull of this dinosaur, by the name of
Majungatholus, was discovered by Dr. Scott
Sampson and his team. As a result of this find
it is now known to be an abelisaurid theropod
similar to *Carnotaurus*.

In recent years Professor Taquet has led
several expeditions into Laos and Cambodia to
sites first worked in the 1940s by the French
geologist J. Hoffet. New specimens of sauropods, hadrosaurs, and theropods have been
collected but they await detailed study.

The institute also has a skeleton of *Edmontosaurus*, and a skull of *Triceratops*, specimens that
were purchased in 1911 from the Canadian
fossil-hunter Charles Sternberg.

225

The Jura Museum, Eichstätt

Holzmaden and Solnhofen

Germany

These late Jurassic sites are famous for the fine state of preservation of their vertebrate fossils. These include *Archaeopteryx*, the rare feathered dinosaur from the Solnhofen deposits, and the complete body fossils at Holzmaden of ichthyosaurs and other marine reptiles.

Holzmaden has yielded the world's best fossils of marine reptiles, especially plesiosaurs and ichthyosaurs. Ichthyosaur fossils showing the outline of the entire animal, including its dorsal fins, have been preserved there. Some fossils, astonishingly, show young inside adult females and there is even a remarkable fossil of an ichthyosaur giving birth. Such fossils occur when mud slides or freak storm events bury the animals without warning. To date, no dinosaurs have been found at Holzmaden, but its fauna includes at least four different types of ichthyosaur, a pliosaur, a plesiosaur, and two species of teleosaurid crocodilians.

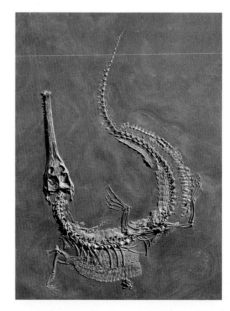

This beautifully preserved Jurassic crocodilian, Steneosaurus, *was found at Holzmaden.*

The Solnhofen deposits are of finely layered lithographic limestones, which outcrop in an east–west belt north of Munich and south of Nuremberg. The main sites are around the towns of Solnhofen, Eichstätt, and Kelheim. They have produced an extremely diverse and well-preserved fossil assemblage of some 360 species of invertebrates, vertebrates, and plants, which inhabited the shallow waters of a lagoon. The bed of this lagoon could not support life, so that animals and plants that fell to the bottom were not scavenged or disrupted but became buried intact.

The vertebrate fauna of Solnhofen includes 54 species of fossil fishes, 28 kinds of reptiles (mostly pterosaurs, but also some ichthyosaurs, plesiosaurs, and crocodilians), and two dinosaurs. These are the complete skeleton of the little coelurosaur *Compsognathus* and the

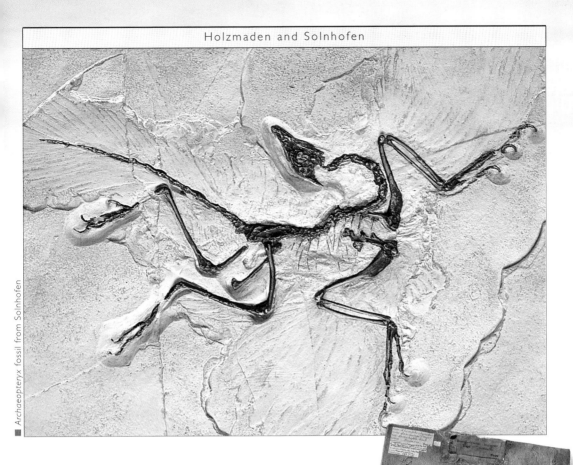

Archaeopteryx fossil from Solnhofen

feathered dinosaur *Archaeopteryx*, now known from about eight specimens in total.

In 1855 the first *Archaeopteryx* discovery—of a partial skeleton without feathers—went largely unnoticed. A feather was found in 1860, which alerted anatomists that an early bird might have existed in the Jurassic. In 1861, a complete skeleton with feathers was found and was purchased by the British Museum of Natural History. This was the subject of Sir Richard Owen's monograph on the anatomy of the first bird. The next specimen was not found until 1877. The most recent finds of *Archaeopteryx* specimens, between 1951 and 1996, have shown the existence of two species—*A. lithographica* and *A. bavarica*. The little dinosaur *Compsognathus* is so similar in its skeletal form to *Archaeopteryx* that one of the first *Archaeopteryx* specimens was for many years mislabeled as a *Compsognathus*, mainly because it lacked feathers.

Solnhofen has also produced many superb pterosaur fossils including *Rhamphorhynchus*, *Scaphognathus*, *Pterodactylus*, *Anurognathus*, *Ctenochasma*, *Gnathosaurus*, and *Germanodactylus*.

FIELD NOTES

Central Europe (Germany)

Late Jurassic

Solnhofen: Archaeopteryx, Compsognathus; Holzmaden: ichthyosaurs, plesiosaurs

Jura Museum, Eichstätt, Germany; museums in Munich and Berlin; Natural History Museum, London, UK; Holzmaden ichthyosaurs in many museums worldwide

Natural exposures can be visited in some areas. Some private sites not accessible to visitors

Pterosaur fossil from the Solnhofen deposits..

Some specimens show exquisite preservation of the delicate wing membranes of these flying reptiles as well as outlines of the tail rudders. Other vertebrate fossils from Solnhofen deposits include rare but very well-preserved ichthyosaurs and crocodilians. Many kinds of fishes and invertebrates were present in the lagoon, and these were what probably lured pterosaurs and small dinosaurs to the water in the first place.

In 1994, a superb book showing the extraordinary riches of the fossil fauna and flora of the Solnhofen sites was published by Dr. Karl Frickhinger.

227

■ Closeup of *Eoalulavis*

Las Hoyas

Spain

The Las Hoyas site in Spain, in the Iberian Mountain Ranges of Cuenca Province, is an ancient lake deposit which has produced some of the world's best preserved fossil birds and rare dinosaurs, as well as a wealth of fossil plants, fishes, amphibians, reptiles, insects, and crustaceans.

The site was discovered in the late 1980s, and the first publication detailing its fossil fauna was released in 1988. The deposit lies within the Calizas de la Huerguina Formation and is dated at the Barremian stage of the early Cretaceous (approximately 117–114 million years old). The fossils are preserved in two kinds of environmental settings—one a lake deposit, the other river sediments marginal to the lake. The detrital river sediments of the formation have yielded less complete remains but nonetheless they make a contribution to our understanding of the entire original fauna and flora of the region. The lake deposits are made up of very fine-grained limestones. This means that carcasses that sank to the bottom of the lake would have been buried quickly, with their organs intact. This kind of preservation shows the feathers on the birds and the outlines of

soft tissue on the other organisms—such as skin impressions for the reptiles, and outlines of horny sheaths over the claw bones of the foot. Furthermore, any bony fossils found in such limestone layers can be prepared out in three-dimensional form using weak acid solutions. Several of the best of the bird specimens, as well as the dinosaur *Pelecanimimus*, were prepared in this manner after first being embedded in a sheet of epoxy resin.

Las Hoyas has yielded thousands of beautifully preserved complete fossil birds. These belong in three genera: *Concornis*, *Iberomesornis*, and *Eoalulavis*. *Iberomesornis* is the most primitive of the three (with regard to its hindlimb and pelvic structure), as it shares certain features with the primitive avian theropod *Archaeopteryx* from the late Jurassic of Germany. Both *Eoalulavis* and *Concornis* belong to the enantiornithine group.

Dr. José Sanz, who is currently in charge of excavations at Las Hoyas, examines the fossil of a hatchling bird.

A fossilized head (above) and feet (below) of Pelecanimimus, *found at Las Hoyas.*

FIELD NOTES

Western Europe (Spain)

■ Late Cretaceous

Pelecanimimus, iguanodontids, bird fossils

Various Spanish museums

■ Not accessible to visitors at present

integumentary fibers as well as a keratinous sheath over the foot claws. The skull is very primitive in retaining a high number of teeth (about 220 in all). This has led to the theory that ornithomimosaurs did not lose their teeth but gradually incorporated many smaller ones into the edges of the beak. The beak was too thin and delicate for slicing and probably would have been used for filter feeding.

Las Hoyas, like Liaoning in China, is one of the world's most important fossil sites for understanding the transition from dinosaurs to birds. Current excavations at the site, which are under the direction of paleontologist Dr. José Sanz, are expected to bring to light new, well-preserved dinosaur remains and more exquisite fossil birds.

These were true flying birds, but they fit in right at the base of the radiation of modern birds. *Eoalulavis* is of particular importance in that the alula is present in its wrist. This bone is necessary for controlled, low-speed flight maneuvers, and crucial for more efficient take-offs and landings.

Only three dinosaurs have been found in the Las Hoyas deposits. Some isolated bones of the ornithopod *Iguanodon* and an as yet undescribed sauropod have come from the detrital river sediments, and a remarkably well-preserved partial skeleton of a primitive ornithomimosaur, *Pelecanimimus*, has come from the fine-grained lake sediments. *Pelecanimimus*, which was about 6–8 feet (2–2.5 m) in length, shows the impressions of skin and

This crocodile fossil, found at Las Hoyas in 1993, dates back to the early Cretaceous.

Institute of Vertebrate Paleontology and Paleoanthropology

Beijing, China

I n 1929 at Zoukoudian Cave near Beijing, a Chinese scientist, Pei Wen-Zhong, discovered an extraordinary fossil skull of a primitive human. This specimen would become world famous as "Peking man," and the discovery provided Chinese authorities with the impetus to start investigating the science of vertebrate paleontology.

The Cenozoic Research Laboratory of the Geological Survey of China thus began to collect vertebrate fossils and in 1957 a separate institution, the Institute of Vertebrate Paleontology, was established. In 1960, its name was changed to the Institute of Vertebrate Paleontology and Paleoanthropology (IVPP), and its scope was widened to include research on human evolution.

The founding director of the IVPP, Professor Yang Zhongjian (1897–1979), became better known in the Western world as Professor C. C. Young. During

FIELD NOTES

🦕 *Chinese dinosaurs, including Lufengosaurus, Monolophosaurus, Psittacosaurus, Protoceratops, Tuojiangosaurus*

the 1950s, Professor Young undertook fieldwork at the Cretaceous dinosaur localities of Shandong Province, where he discovered remains of the unusual hadrosaurid *Tsintaosaurus* as well as skeletons of psittacosaurids. Young also studied the giant sauropod *Mamenchisaurus*, an almost complete skeleton of which was collected in Szechuan in 1957. Between 1963 and 1966, Dr. (now Professor) Dong Zhiming began exploring the Jungaar and Turpan basins of northern China. Here he discovered a fauna that included the dinosaurs *Monolophosaurus*, *Tienshanosaurus*, *Bellusaurus*, *Psittacosaurus*, *Wuerhosaurus*, *Jaxartosaurus*, and *Kelmayisaurus*. Throughout the 1970s and 1980s, teams from the IVPP, often working

In 1993 the IVPP moved into this modern building in Beijing.

A scientist prepares an exhibit at the IVPP

in conjunction with local scientific authorities from the various provinces of China, made many exciting dinosaur discoveries.

Professor Dong has probably discovered and named more dinosaurs than any other 20th-century paleontologist. Perhaps one of his most important discoveries was the Dashanpu Dinosaur Quarry near Zigong in Szechuan Province. Today, an impressive museum has been built above the site to house many of its important specimens as well as to show the excavations in progress. More than 8,000 fossil bones, including many complete dinosaur skeletons, have been excavated or exposed from the site.

During the late 1980s and early 1990s, joint expeditions of the IVPP and the Royal Tyrrell Museum, Canada, explored the remote regions of northern China and inner Mongolia and discovered numerous new dinosaurs and other vertebrates. Many of the specimens that they found, such as the theropod *Sinraptor*, are now in the IVPP collections.

The IVPP conducts research on all aspects of vertebrate paleontology and paleoanthropology and curates a huge collection of fossil specimens. Each year its field parties scour the far corners of China in their search for fossils. The IVPP is divided into three departments—paleoichthyology and paleoherpetology (fishes, amphibians,

Professor Dong Zhiming, one of China's most celebrated paleontologists, is a prolific discoverer of new dinosaurs.

reptiles, and dinosaurs), paleomammalogy (fossil mammals), and paleoanthropology (human fossils). In the early 1990s, the IVPP had a full-time staff of 230 people. Results of the IVPP's research work are published through its three scientific journals as well as through major international scientific journals. The IVPP holds over 200,000 specimens of fossil vertebrates. These include most of the type or reference specimens for many Chinese dinosaurs.

In 1993 the IVPP moved into a new, larger building which has a three-story public area for fossil displays. In the last few years, the IVPP has collected a number of extraordinary remains of feathered Chinese dinosaurs and primitive birds from the early Cretaceous Liaoning sites in northern China.

231

Zigong Dinosaur Park

Szechuan, China

The Dinosaur Museum in Dashanpu, in Szechuan Province, was opened in the spring of 1987.

The Zigong Dinosaur Park, in central Szechuan Province, near Dashanpu, Zigong, opened in 1987. It was erected over the top of a rich site that contained many dinosaur skeletons alongside other fossil vertebrates. The fossils occur in gray-green sandstones and mudstones of the middle Jurassic Shaximiao Formation, an ancient river setting. Dinosaurs and other animals probably drowned in flood events and their bodies would have accumulated at river bends where the current slowed down. In this respect the Dashanpu site is the Chinese equivalent of the rich bone bed at Dinosaur National Monument in Utah.

The Zigong Dinosaur Museum was the first museum dedicated solely to dinosaurs to open in Asia. Today it features the most impressive display of Chinese dinosaur skeletons in any one place, with an excavation of real bones exposed for public viewing in the museum's bedrock. Fossils were first found near Dashanpu in 1972 by geologists from the Szechuan Bureau of Geology and Mineral Resources—they came upon a dinosaur bone sticking out of a road-side cliff. The site was first excavated in 1977 and produced a large sauropod skeleton. It was named by Professor Dong Zhiming and his colleagues as *Shunosaurus*. On December 17, 1979, Professor Dong and Zhou Shiwu discovered many more dinosaur remains, exposed in situ in the Dashanpu region. Excavations at the site from 1979 to 1981 produced further skeletons of *Shunosaurus*, as well as the theropod *Gasosaurus*, a beautifully preserved stegosaur *Huayangosaurus,* and a new sauropod named *Dataousaurus*. Further excavations the following year by He Xinlu of the Chen-Du College of Geology uncovered skeletons of small ornithopods at the site, together with a large, new sauropod, *Omeisaurus*.

In 1984, while the museum was being built, almost 5,400 square feet (500 m²) of bedrock had to be cleared. This resulted in more fossil discoveries, such as a skull of a late-surviving

The center of the museum's main hall

Local villagers watch the activities of a group of scientists working on dinosaur fossils in the Zigong Dinosaur Park.

skeletons. These include the sauropods *Omeiosaurus* and *Shunosaurus*, the stegosaur *Huayangosaurus*, the theropod *Gasosaurus*, and the small ornithopod *Xioasaurus*. The most spectacular part of the display is the exposed bedrock in the center of the main hall, which shows a large number of dinosaur and other vertebrate bones just as they were found.

Geological investigations at Dashanpu have shown that the site has enormous potential for more discoveries. Today the excavated area covers some 31,000 square feet (2,800 sq m) but the bone bed continues for an estimated 215,000 square feet (20,000 sq m). To date, over 8,000 bones of dinosaurs and other vertebrates have been excavated from the site. Research on the Dashanpu fossils is ongoing—excavations are revealing more bones every year.

labyrinthodont amphibian, *Sinobrachyops*, and the skull of a mammal-like reptile, *Bienotheroides*. Remains of freshwater fossil fishes, turtles, crocodiles, a plesiosaur, and a pterosaur have also been excavated from the site. The pterosaur, *Angustinaripterus*, was a close relative of *Rhamphorhynchus* from Germany. The total number of vertebrates collected from the Dashanpu Quarry is now 23 species, eight of which are dinosaurs.

The museum's basement is used for fossil storage and fossil preparation and there are two floors of displays. Displayed in the main hall are 10 almost complete, mounted dinosaur

FIELD NOTES

🧭 Central China (Szechuan)

◾ Late Jurassic

🦖 Large sauropods and theropods (e.g. Shunosaurus, Omeiosaurus, Yangchuanosaurus)

🏛 Zigong Dinosaur Museum, Dashanpu (on site)

◾ Open to visitors

233

Storm clouds hover over Liaoning Province

Liaoning

China

Quarries near Sihetun in the province of Liaoning, northern China, have recently become famous, due to their remarkable preservation of primitive birds and small dinosaurs, with feathers and integumentary coats preserved. The animals drowned in a lake and sank to the muddy bottom, where they were rapidly buried by nearby volcanic ash falls without any disruption of their remains. Today their fossils occur in fine-grained, layered siltstones, which can be split with chisels to reveal the specimens. The site is of uncertain age, some estimates placing it as latest Jurassic, but recent work supports an early Cretaceous date (around 140–120 million years ago). The bird fossils are especially significant in being the first record in geological time of birds after the famous late Jurassic *Archaeopteryx* specimens from Germany.

The discovery of the little dinosaur *Sinosauropteryx* in 1996 caused a sensation because it was the first dinosaur to be found showing an integumentary coat of what appeared to be fiberlike feathers. Further analysis has shown that these are primitive feathers, but they lack the complex bifurcations seen in the feathers of flying birds. Other soft-tissue preservation in the dinosaurs of this site show the remains of gut contents, eggs inside oviducts, and impressions of soft organs, such as the eye capsule.

Paleontologists from the Institute of Vertebrate Paleontology and Paleoanthropology in Beijing excavate layers of sedimentary rock at Jianshangou in Liaoning Province.

To date, several kinds of theropod dinosaurs have been excavated from Liaoning. All show feathers or fibrous coats over parts of the body. *Sinosauropteryx*, the most primitive of these theropods, is a small coelurosaur that is similar to *Compsognathus*. *Sinornithosaurus*, a primitive dromaeosaurid, and *Beipiaosaurus*, a therizinosaur, also have simple fibrous feather coats, similar to that seen in *Sinosauropteryx*. Two small dinosaurs, however, are extraordinary in that they have longer feathers with bifurcating broad

234

Paleontologists, with the help of local farmers, excavate near Sihetun

areas, as evident in typical bird feathers. These are *Caudipteryx* and *Protoarchaeopteryx*. Both dinosaurs have longer arms and shorter legs than the other dinosaurs of Liaoning, and clearly represent a more advanced step toward the evolution of birds. They still lack arms large enough for them to be functional flying wings, but their overall general appearance does resemble that of birds more than it does dinosaurs.

Many thousands of very beautifully preserved primitive birds are known from the Liaoning site, the most primitive species being *Concornis* and *Confuciusornis*. Both of these are toothed birds, slightly more advanced in their skeleton than *Archaeopteryx*. If *Archaeopteryx* is considered to be a feathered dinosaur, then they probably represent the first true birds.

No site on Earth has done more to reveal the many stages of evolution from agile, running dinosaurs to flying birds than Liaoning. A large on-site museum has recently been built to oversee further excavations, house the collections,

FIELD NOTES

Northern China

Early Cretaceous

Feathered dinosaurs (e.g. Sinosauropteryx, Sinornithosaurus), Psittacosaurus; birds

On-site museum; IVPP, Beijing; China

On-site museum open to visitors

and display the most impressive specimens for vistors. It will also serve as an important research facility for scientists. Since the first feathered dinosaur was discovered there, scientists from the Institute of Vertebrate Paleontology and Paleoanthropology in Beijing have been working closely with paleontologists from Canada, the United States, and European countries to study the new discoveries that continue to come out of this rich and exciting site.

Li Yinfang, the farmer who in 1996 discovered the first fossil specimen of Sinosauropteryx.

235

The Flaming Cliffs of Xinjiang

Flaming Cliffs

Mongolia

This renowned site in the remote pre-Altaic Mongolia was discovered in 1922 by the famous Central Asiatic Expedition into the Gobi Desert, led by Dr. Roy Chapman Andrews of the American Museum of Natural History (AMNH). The expedition became lost and stopped to ask directions from a local Mongolian settlement. While Andrews went to talk to the locals, the expedition photographer, J.B. Schackelford, wandered off toward some nearby red rocks to take photos. Within minutes he was standing on a huge cliff, at the base of which he started to find fossil bones. He quickly alerted his colleagues, and by the end of that day they had found the first ever dinosaur nest and a skull of *Protoceratops*. The expedition had little time to explore. However, when they returned in 1923, the researchers found a complete nest of eggs they thought were those of *Protoceratops*, but are now known to be those of *Oviraptor*, and skeletons of articulated dinosaurs, including the theropods *Velociraptor*, *Sinornithoides*, and *Oviraptor*, the ankylosaur *Pinacosaurus*, and abundant remains of the neoceratopsian *Protoceratops*.

The expedition gave the name Flaming Cliffs to the uppermost part of a 6-mile (10 km) escarpment of late Cretaceous sandstones known as Bayn Dzak. The rocks outcropping there belong to the Djadokhta Formation and are composed in the main of poorly cemented dune and alluvial plain sandstones dated at Campanian stage (around 87–72 million years old). Bayn Dzak is now the designated type section for the geology of the Djadokhta Formation.

In 1948 a Russian expedition was made to Flaming Cliffs, and between 1963 and 1971 Polish–Mongolian teams, led by Professor Zofia Kielan-Jaworowska visited the site a number of

Members of the 1922 American expedition to the Gobi Desert on a cliff top in the badlands at Urtyn Obo.

A camel caravan entering the valley of Shabarakh Usa

times. In 1965 one of the Polish–Mongolian expeditions, comprising some 23 participants, discovered a nest of juvenile *Pinacosaurus* skeletons and more dinosaur eggs at the site.

The expeditions that followed over the next few years concentrated on searching for small vertebrate fauna at Bayn Dzak. They discovered many species of lizards, crocodiles, and small mammals. Other discoveries in the Djadokhta Formation included articulated skeletons of *Protoceratops* in upright positions, suggesting that they were buried suddenly in sandslides.

Perhaps their most famous find from the region was made in 1971 at a site called Toogreeg, 50 miles (30 km) from Bayn Dzak. This was two complete, interlocked dinosaur skeletons, the arms of the *Velociraptor* gripping the skull of the *Protoceratops*, suggesting they were fighting when suddenly buried by a freak sandstorm.

Since 1990, expeditions led by Dr. Mike Norell with Dr. Mark Norell of the AMNH, have again explored the Djadokhta Formation outcrops and made many finds from new localities such as Ukhaa Tolgod, which was dis-

Members of the American and Mongolian field crews search for dinosaur fossils in the Gobi Desert in 1991.

covered in 1993. These finds include the first articulated skeletons of oviraptorids brooding over their nests of eggs, some of which contain embryonic dinosaur skeletons. The site's geology suggests that when these dinosaurs lived it was a hot, arid environment with irregular rainfall. The site seems to have preserved animals like snapshots in time as huge sand and dust storms buried them rapidly, without warning. Ukhaa Tolgod is, according to Dr. Mark Norell, the richest Cretaceous site yet found in Asia, possibly the world, for its abundance of well-preserved vertebrate fossils.

FIELD NOTES

✴ Gobi Desert, Mongolia

◼ Late Cretaceous

🦅 Large and small theropods (e.g. Tarbosaurus, Jenghizkhan, Velociraptor, Therizinosaurus), sauropods, hadrosaurs, neoceratopsians

🏛 IVPP, Beijing, China, Academy of Science, Ulan Baatar, Mongolia; American Museum of Natural History, New York, USA;

◼ Access very difficult

237

Atacama Desert, Valley of the Moon

Valley of the Moon

Argentina

The Valley of the Moon (Valle de la Luna) is a region in the north-west of Argentina in the Province of La Rioja, where there is a thick succession of middle to late Triassic sedimentary rock outcrops. The Valley of the Moon and the region that surrounds it are famous for their Triassic vertebrate localities. These contain the oldest well-preserved dinosaur fossils anywhere in the world as well as a range of other more primitive archosaurian reptiles that show evolutionary links to the first dinosaurs.

The earliest dinosaurs come from the famous Ischigualasto Formation. This formation is a succession of sandstones, mudstones, and siltstones that have weathered to form the strange, alien-looking landscape of deeply eroded gullies and steep cliffs for which the valley is named. The fossils of the Ischigualasto Formation come from outcrops in the foothills of the Andes, and are dated within the Triassic period as belonging to the Carnian stage, between approximately 226 and 220 million years ago.

Dinosaurs from the region are very sparse, and intensive fieldwork was required to find the few specimens that have been discovered there.

FIELD NOTES

☀ *Southern South America*

◼ *Late Triassic*

🦅 *Herrerasaurus, Staurikosaurus, Eoraptor, other archosaurians*

🏛 *Buenos Aires Natural History Museum, Argentina*

◼ *Remote location, but accessible*

The first dinosaurs from the region were discovered by Dr. Osvaldo Reig between 1959 and 1961. In 1960 he found the bones of the basal saurischians *Herrerasaurus* and *Ischisaurus* and he formally described them in 1963. Soon after that, the first primitive ornithischian dinosaur—based on some jaw fragments, a few leg bones, and some vertebrae—was discovered here. It was described by Dr. R. Casamquela in 1967 as *Pisanosaurus mertii*.

During the 1988 field season, a combined team of paleontologists from both North and South America made many discoveries in the Valley of the Moon and its surrounding sites. More complete material of the basal dinosaur *Herrerasaurus* was uncovered, and these new finds have made *Herrerasaurus* one of the most fully described dinosaurs from the region. At 15 feet (4.5 m) long, it was the largest predatory dinosaur known from the late Triassic.

As well as *Herrerasaurus*, the team discovered an almost complete skeleton of a new, very primitive little carnivorous dinosaur. Dr. Paul Sereno and his colleagues named it *Eoraptor lunensis*. *Eoraptor* was only about 3 feet (1 m)

Dr. Osvaldo Reig, who first discovered dinosaurs in the Ischigualasto Formation.

long and its skull displayed almost all the basic traits of a dinosaur. It is still a controversial fossil: Some scientists believe *Eoraptor* to be the oldest true dinosaur, whereas others think it may be too primitive to be considered a dinosaur. The debate hinges on the arbitrary criteria that are used to define the dinosaur group.

The Valley of the Moon has produced other interesting animals in addition to its dinosaurs. *Sillosuchus*, a predatory archosaur, was discovered in the region in 1979 by paleontologists from the National University of San Juan. Remains of a juvenile rhynchosaur, *Scaphonyx*, were also discovered inside the ribs of one *Herrerasaurus* specimen, providing valuable direct evidence of the diet of some of these early dinosaurs.

Underlying the Ischigualasto Formation is the Los Chanares Formation, which has yielded

the skeletons of advanced archosaurs that are considered to be immediately ancestral to the first dinosaurs. These beasts—*Lagosuchus*, *Lagerpeton*, and *Pseudolagosuchus*—lived between about 230 and 225 million years ago. They were the first reptiles to develop upright running postures.

Most of the dinosaurs and archosaurians found from the Valley of the Moon are housed in the collections of the Buenos Aires Natural History Museum.

Herrerasaurus was named after a goat farmer, Victorino Herrera, who discovered the first skeleton of this dinosaur in 1963.

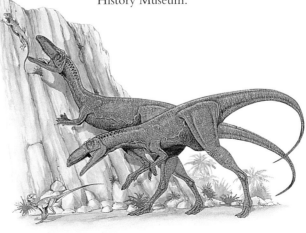

239

Karoo Basin

South Africa

The Karoo Basin covers almost two-thirds of the surface area of South Africa. It is a thick sequence of sedimentary rocks laid down by large river systems, ranging in age from the middle Permian through to the early Jurassic (about 190 million years ago). The Karoo rocks are subdivided into a number of sequences, of which the Beaufort Group (middle Permian–middle Triassic) has yielded most of the important specimens of amphibians, reptiles, and mammal-like reptiles. The Stormberg Group (late Triassic–early Jurassic) is situated above the Beaufort Group and contains examples of the oldest dinosaurs in southern Africa.

The Beaufort Group has been subdivided into a number of biostratigraphic zones based on the fossil assemblages they contain. The significance of the Karoo Basin sequence is that its 50-million-year fossil record is largely unbroken, so that the evolutionary patterns of the different animal groups can be followed through time.

About 240 million years ago, in the middle Permian, South Africa's first terrestrial reptiles evolved. Some were gigantic meat-eaters, some were large plant-eaters,

and others were small, active predators. The fossilized skeletons of these animals are often found complete and undisturbed in Karoo sediments, and it is possible that many died after becoming trapped in the soft mud in which their bones were preserved. The synapsids, or mammal-like reptiles, were the dominant animals of the early Karoo. The first mammals evolved from these early in the late Triassic. Among the most spectacular of the Karoo reptile finds were the dinocephalians. Some of these large animals, such as *Anteosaurus*, were meat-eaters; others, such as *Moschops*, were herbivores.

The many layers of sedimentary rocks in the Karoo Basin have each yielded distinct faunas (thus enabling geologists to correlate the rock sequences across the African continent), and serve as a basis for correlations right across parts of Gondwana. Such zones are based on the most common animals that have been uncovered in each layer and they are named accordingly—for example, *Cynognathus* zone or the *Lystrosaurus* zone. Since the

Andrew Geddes Bain, the Scottish engineer who found the first fossil reptiles in the Karoo.

240

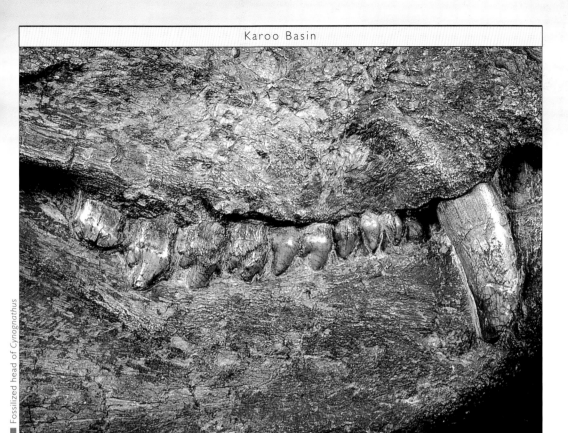

■ Fossilized head of *Cynognathus*

first fossil reptile find in 1838 by Andrew Geddes Bain, several of these Triassic reptile fossils from the Karoo Basin have turned up in other Gondwana regions. *Lystrosaurus*, for example, is now known also from India, China, and Antarctica.

Rocks of the Karoo zones are exposed in several classic localities throughout South Africa. The *Tapinocephalus* zone, the earliest part of the Karoo succession, is exposed in the area lying south of Beaufort West. The *Endothiodon*, *Cistecephalus*, and *Dicyodon* zones have outcrops throughout the Beaufort West district. Fossils of the *Lystrosaurus* zone, which mark the base of the Triassic Period, are found near Middelburg (Cape), Bethulie, Bergville, Bloemfontein, and Harrismith. The classic *Cynognathus* zone localities are Burghersdorp and Aliwal North, while the red beds of the Molteno and Elliot formations surround the Drakensberg Mountains and can be seen in the Herschel, Matatiele, and Fouriesburg districts.

FIELD NOTES

✳ South Africa

■ Middle Permian–early Jurassic

🦅 Massospondylus, Euskeleosaurus, Syntarsus

🏛 South African Museum, Cape Town; Bernard Price Institute, Witwatersrand University, Johannesberg, South Africa; various other South African museums

■ Most sites on private land; accessible only with owners' permission

The most notable sites for dinosaur and reptile fossils in the Karoo Basin are in the early Jurassic Elliott Formation and Bushveld Sandstone. Here, well-preserved remains of a number of early dinosaurs, such as the prosauropods *Melanorosaurus*, *Massospondylus*, and *Euskelosaurus*, have been found, together with the basal ornithischians *Heterodontosaurus* and *Fabrosaurus*, the small ceratosaur *Syntarsus*, and the large, predatory archosaurian reptiles such as *Euparkeria* and *Chasmatosaurus*. Theropod footprints have also been found in the Molteno Formation of the Karoo Basin

The Karoo Basin shows the most detailed record of the transition of life on land across the Permian–Triassic boundary, which, like the later Cretaceous–Tertiary boundary, was one of the Earth's major global extinction events.

An example of theropod footprints found in the Molteno Formation of the Karoo Basin.

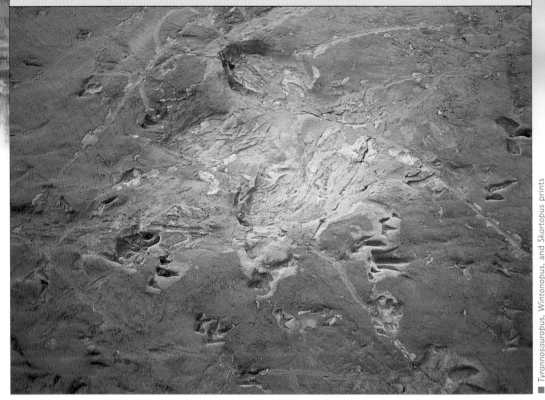

Tyrannosauropus, Wintonopus, and Skartopus prints

Lark Quarry

Queensland, Australia

Lark Quarry, near Winton, Queensland, in north-eastern Australia, is one of the best preserved dinosaur trackway localities in the world. The site documents the activities of three different species of dinosaurs. The footprints are preserved in a fine-grained sandstone which is known as the Winton Formation and is dated at between about 95 and 90 million years old. More than 3,300 footprints, representing some 150 individual dinosaurs, tell the story of what took place in about 10 seconds of time. The footprints are now on public display and protected by an enclosure at Lark Quarry on the road outside of Winton.

A herd of small, grazing plant-eaters (which have been given the footprint type name *Wintonopus*) and a large number of small, predatory coelurosaurs (footprint type name *Skartopus*) were disturbed by a large carnivorous dinosaur (footprint type

name *Tyrannosauropus*), which is thought to have cornered them against some rocky bluffs. The smaller dinosaurs had no choice but to run out past the meat-eater, their tracks indicating that they sprinted at top speed—about 12 miles per hour (20 km/h) judging by the length of their strides—to get around their enemy. The dinosaur footprints were studied by Drs. Tony Thulborn and Mary Wade, who published their results in 1984 in the *Memoirs* of the Queensland Museum.

At the Lark Quarry site, there are more than 1,000 well-preserved footprints that represent *Wintonopus*. These prints are recognizable by their asymmetry, rounded toes and lack of a heel impression. Thulborn and Wade suggested that the animal that made these

The sandstone of the Winton Formation has preserved thousands of dinosaur footprints.

Landscape near Winton, Queensland

prints was a medium-sized ornithopod. Of 57 tracks analyzed, only one appeared to have been made by an animal larger than 3 feet (1 m) high at the hip; in fact, 81 percent of them seemed to be less than 20 inches (50 cm) high at the hip. The animals moved at a fast, running gait at about 7–12 miles per hour (12–20 km/h). The tracks show no evidence that the animals were slowing down.

Skartopus has left sharp, three-toed prints, which are almost symmetrical and range in length from a little over 1 inch (2.9 cm) to 2¼ inches (5.7 cm). The animal that made these tracks was probably a small coelurosaur (in the size range between "bantams and half-grown emus," according to Thulborn and Wade), and was moving at a mean speed of 8 miles per hour (13 km/h). A measured sample of 34 of the tracks concluded that the animals represented a range of normal size variation, and that they had a maximum hip height of about 8½ inches (22 cm). The fact that the *Skartopus* tracks are

One of the 11 footprints of Tyrannosauropus that have been preserved at Lark Quarry.

running in the direction opposite to those of the large carnivore suggests that there was a herd of these animals trying to escape the predator.

Tyrannosauropus's trackway is represented by 11 footprints, some of which show evidence of a sharp, pointed claw on the end of each toe. The prints have a mean length of 20½ inches (52 cm), suggesting the predator was 30–33 feet (9–10 m) long. Unlike those of *Megalosauropus*, the largest of which are of similar length, the *Tyrannosauropus* prints are broader and deeper, suggestive of a heavier animal—probably similar to *Tyrannosaurus*. The stride of the large predator ranged from 9 to 12 feet (2.8-3.7 m) and it moved at about 3–5 miles per hour (5–8 km/h)—a walking gait. It seems to have taken a slightly weaving course, with a tendency to slow down, as the first four strides are longer and deeper than the next four.

FIELD NOTES

✦ North-eastern Australia

■ Early late Cretaceous

🐾 Superb trackways of Skartopus, Tyrannosauropus, and Wintonopus

🏛 On-site display

■ Accessible to visitors

243

■ Dinosaur Cove

Dinosaur Cove and East Gippsland

Victoria, Australia

A small claw of a theropod dinosaur, found in 1904 by the geologist William Ferguson, was the first bone to be recovered from the early Cretaceous coastal rocks of Victoria, Australia. In 1978 three fossil-hunters—Tim Flannery, John Long, and Rob Glenie—went again to the Cape Patterson site. Almost immediately, Long found another dinosaur bone, and over the next few weeks, more bones were collected. As a result, a thorough search of the area was undertaken, led by Dr. Tom Rich of the Museum of Victoria. Although Dr. Rich, and his wife, Dr. Pat Vickers-Rich, originally focused on finding fossil mammals, their search was to yield many important dino-saur discoveries. In 1980, Tim Flannery found a theropod anklebone (astragalus) near Inverloch. This was published in 1981 as belonging to the theropod *Allosaurus*, a genus then known only from the much older Jurassic of North America.

In the early 1980s teams led by Tom Rich and Pat Vickers-Rich discovered a site in the Otway Ranges of western Victoria, which was named Dinosaur Cove. Here they found first

FIELD NOTES

✲ South-eastern Australia

■ Early Cretaceous

🦆 Hypsilophodontids
(e.g. Leaellynausaura,
Atlascopcosaurus), rare theropods,
large labyrinthodonts and other
vertebrates

🏛 Museum Victoria, Melbourne,
Australia; Monash University Science
Center, Victoria, Australia; Western
Australian Museum, Perth, Australia

■ Most sites accessible, but some are
dangerous to reach

fossils—a skull and a partial skeleton—of the hypsilopho-dontid dinosaur *Leaellynasaura*. The excavations at the site in-volved tunneling directly into the rugged sea cliffs using mining equipment—thus the world's first dinosaur mine was created. The combined sites of east and west Victoria have so far yielded a diverse assemblage of vertebrates—dinosaurs, plesiosaurs, pterosaurs, amphi-bians, fish, and turtles. Most recently, primitive mammals have also been added to the list. Tom Rich and Pat Vickers-Rich have now named three hypsilophodontids from the sites in Victoria—*Leaellynasaura*, *Atlascopscosaurus*, and *Qantassaurus*—and yet others await description. From the slightly older (Aptian) east Gippsland sites, a jaw and other bones belonging to a giant labyrinthodont amphibian, *Koolasuchus*, were found. It was around 10–13 feet (3– 4m) long and dominated the ancient river system.

In 1994, two very strange dinosaur finds—the bones of an ornithomimosaur and a possible neoceratopsian—were made at the Dinosaur Cove site. Up until then, both of these groups

The "dinosaur mine" at Dinosaur Cove

Part of the rugged coastline of the east Gippsland region in south-eastern Australia.

had been thought to be unique to the late Cretaceous of the Northern Hemisphere. In 1986, a possible oviraptorosaur was described from a jaw fragment and backbone. All of these finds suggest that these dinosaur groups could have had an earlier origin in Gondwana, before they migrated north to Europe, Asia, and North America.

The aspect of greatest significance regarding the Victorian dinosaur sites is their paleogeographic location. At approximately 75° South, the animals would have inhabited a polar forest and endured temperatures below freezing during the winter months. They would

have survived in total darkness for three months of the year. The fact that these dinosaurs managed to survive in these conditions makes them crucial to the ongoing debate about warm-bloodedness in dinosaurs. Studies of their bone histology through thin-sectioning have revealed that some of the theropods, such as *Timimus*, would have had periods of non-growth. This suggests that they hibernated through winter. The hypsilophodonts, on the other hand, show continuous bone growth, so they were probably active through the winter months.

The first mammals' jaws were discovered at the east Gippsland site in 1996, and include the earliest possible placental mammals from Australia—*Ausktribosphenos* and *Teinolophas*—which possibly belonged to the Eupantothere group of primitive mammals.

Footprints of the small theropod Skartopus, one of a number of rare theropods found at Dinosaur Cove,

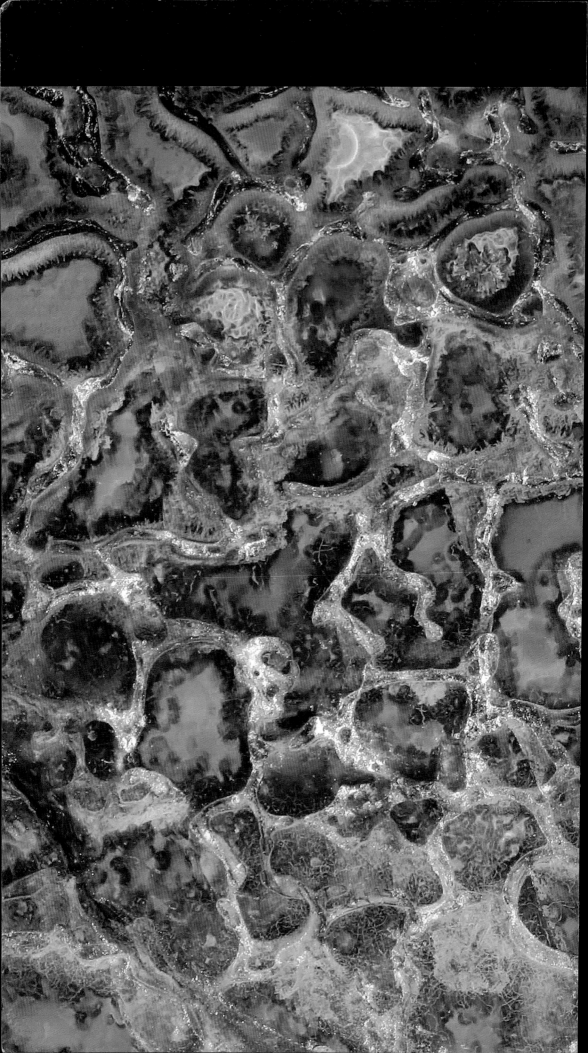

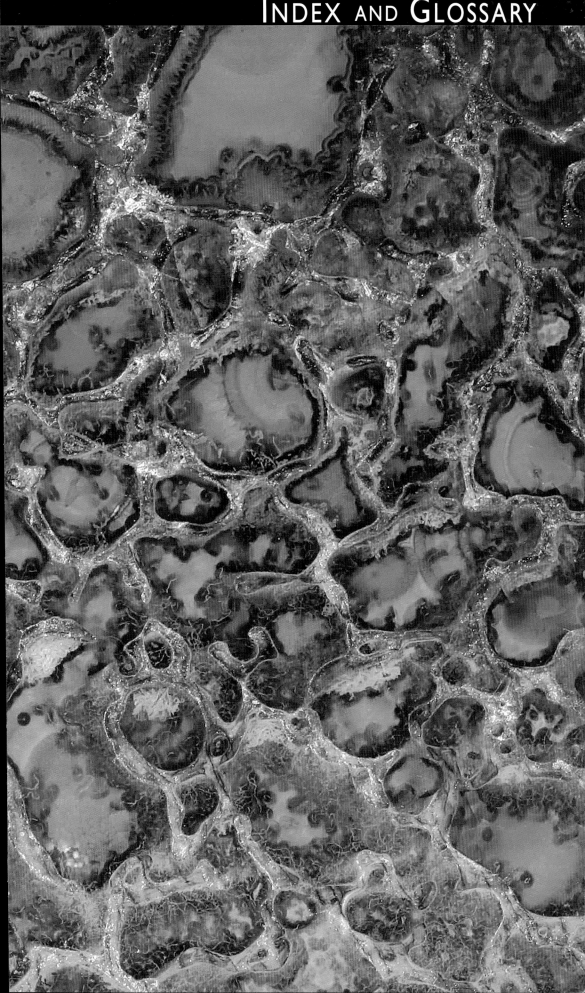

INDEX AND GLOSSARY

In this combined index and glossary, **bold** page numbers indicate the main reference, and *italics* indicate illustrations and photographs.

CONTRIBUTORS

Michael K. Brett-Surman, Ph.D. (consultant), is the museum specialist for dinosaurs at the National Museum of Natural History, Smithsonian Institution, Washington DC, and an adjunct associate professor at George Washington University. He is the author/editor of two award-winning books on dinosaurs and a reviewer/consultant for scientific and popular book publishers. He resides in Virginia.

Christopher A. Brochu obtained his B.S. in geology at the University of Iowa and his Ph.D. in vertebrate paleontology at the University of Texas. Since then, he has been a research scientist at the Field Museum in Chicago, working on a full description of *Tyrannosaurus rex* based on "Sue," the most complete skeleton of this species thus found. His overall research focuses on developing phylogenetic answers for historical questions, especially with crocodiles and dinosaurs. He will be an assistant professor of geoscience at the University of Iowa starting in 2001.

Colin McHenry is the managing curator of the Age of Fishes Museum in Canowindra, New South Wales, Australia.

John Long, Ph.D., is Curator of Vertebrate Paleontology at the Western Australian Museum. His research has focused on the early evolution of fishes and Mesozoic reptiles. He has published more than 100 scientific papers and over 60 popular articles, and is the author of four books.

John D. Scanlon, Ph.D., lives in Townsville, north Queensland. A research associate of the Australian Museum, Sydney, John has also worked at Monash University, Melbourne, the University of Queensland, Brisbane, and the Friedrich-Wilhelms Universität, Bonn, Germany. He works mainly on Australian snakes, including the systematics and behavior of living species as well as fossils.

Paul Willis, Ph.D., has been associated with numerous important fossil excavations. His enjoyment for communicating science to popular audiences has involved him in touring Australian elementary schools with a life-size inflatable *Tyrannosaurus rex* as his lecture companion. He is currently a science reporter with the Australian Broadcasting Corporation. He lives in Sydney, Australia.

CAPTIONS

Page 1: A reconstructed plesiosaur fossil towers over a human observer.

Page 2: This footprint made by a theropod dinosaur has been recorded in stone.

Page 3: Archaeopteryx is a classic and rare example of an evolutionary path between dinosaurs and birds.

Pages 4–5: The entrance hall of the American Museum of Natural History in New York features the skeletons of dinosaurs in battle.

Pages 6–7: Sedimentary rock layers with dinosaur fossil, San Francisco, California, USA

Pages 10–11: Scientists at the Sahat Sakan Paleontology Institute in China study the skeleton of a young sauropod.

Pages 12–13: Excavating dinosaur fossil bones, Dinosaur National Monument, Utah, USA.

Pages 32–3: Moss-covered trees evoke the environment in which the dinosaurs lived and died.

Pages 60–1: A fossil of an *Ichthyosaurus*—it survived two great extinctions only to perish before the dinosaurs 66 million years ago.

Pages 84–5: Dinosaurs have proved to be of perennial fascination to adults and children alike.

Pages 112–3: Researchers brave the elements to examine snow-covered dinosaur tracks.

Pages 206–7: The harsh badland environment yields up fossil evidence of the age of dinosaurs.

Pages 246–7: Agatized dinosaur bone from Morrison formation, Colorado, USA.

ACKNOWLEDGMENTS

The publishers wish to thank the following people for their assistance in the production of this book:
Peta Gorman and Marney Richardson (editorial assistance), Stuart McVicar (banding for chapters 5 and 6), Nancy Sibtain (index), Bronwyn Sweeney (editorial development of synopsis).

PHOTOGRAPH AND ILLUSTRATION CREDITS

(t=top, b=bottom, l=left, r=rght, c=center)
AMNH=American Museum of Natural History, Courtesy Dept of Library Services; ARDEA=ARDEA, London; AUS=AUSCAPE; BAL= The Bridgeman Art Library; BC=Bruce Coleman Limited; IVPP=Institute of Vertebrate Paleontology and Paleoanthropology, Beijing, China; MEPL =Mary Evans Picture Library; NCMNS=North Carolina Museum of Natural Sciences; NHM=The Natural History Museum, London; NGS=National Geographic Society; OLM=O. Louis Mazzatenta; OSF=Oxford Scientific Films; PEP=Planet Earth Pictures; RTMP=Royal Tyrrell Museum of Paleontology; SPL=Science Photo Library; TSA=Tom Stack & Associates; WO=Weldon Owen Pty Ltd.

Photograph Credits
1Getty Images 2 SPL 3 Corbis 4–5 Corbis 6–7 Ken Lucas/California Academy of Science/PEP 8–9 Daniel W. Gotshall/PEP 10–11 Corbis 12–13 Corbis 14 Peale Museum, PA, USA/BAL 15t Tui De Roy/AUS 16tl K. Svensson/SPL/photolibrary.com; b Colin Weston/PEP 17t AMNH (Neg No. 128003) 18tr P. Morris/ARDEA; l Sinclair Stammers/SPL/photolibrary.com; b NHM, 19t Tom Till/AUS 20c Mary Evans-Explorer/AUS 21t OLM; b Reg Morrison/AUS 22tl MEPL; bl Jan Tove Johansson/PEP 23cl SPL/photolibrary.com; br F. Gohier/ ARDEA 24c F. Gohier/AUS 25t Tom & Therisa Stack/TSA; bl F. Gohier/ARDEA; br F. Gohier/AUS 26t Paul Willis; b IVPP 27tl Tom & Therisa Stack/TSA; br Stuart Bowey/Ad-Libitum 28tr Arnaud Fevrier/Royal Belgian Museum, Brussels; bl MEPL 29 NHM, 30tr North Wind Picture Archives; b Peabody Museum of natural History, Yale University 31t AMNH (Neg. No. 411007); cr AMNH (Neg. No. 17808) 32–33 photolibrary.com 35t T. Fedak & A. Letourneau/Fundy Geological Museum 37t G. R. "Dick" Roberts Photo Library, NZ 39b Ron Lilley/BC 40t Tom & Therisa Stack/TSA 43t MEPL 45cr Tom & Therisa Stack/AUS; cl NHM 46c Kevin Aitken/Peter Arnold/AUS 47t Queensland Museum; cl P. Morris/ARDEA 48b John Cancalosi/BC 49tl Ferrero-Labat/AUS; tr AMNH (Anderson/Neg. No. 35486) 50tr NASA; b Jean-Paul Ferrero/AUS 51t F. Gohier/ ARDEA; c NASA; b Paul Franklin/OSF 52 C. Pouedras/ Eurelios/SPL/photolibrary.com 53b Tui De Roy/AUS 54tl Corel Corp. 55t Paul Thompson/ photolibrary.com; c Corel Corp. 56b OLM 57t Dr Scott Nielsen/BC 58b Peabody Museum of Natural History, Yale University 59tl Jane Burton/BC; tr OLM; b NHM 60–61 Corbis 62tl AMNH (Neg. No. 35044) 63 Corel Corp. 64tl AMNH (Anderson/Neg No. 35423); tr AMNH (Neg. No. 325277) 65c Mark A. Philbrick/Brigham Young University 66t NCMNS 67b Tom and Pat Rich 68t Chris A. Brochu 69t NCMNS; br NHM 70tl NHM; b David G. Gillette 71t NHM; c OLM 73t Hulton Deutsch/ photolibrary.com 74tl Tom & Therisa Stack/TSA; tr AMNH (Ellison/No. K17685); b Xinnua-China/GAMMA/ Picture Media 75b NHM 77t AMNH (Neg. No. 338623) 78tr NHM 79t Zofia Kielan-Jaworowska; br Paul Willis 81t NHM; c F. Gohier/ ARDEA 82b F. Gohier/ARDEA 83 NHM 84–85 Corbis 86tl National Philatelic Collection, Australia Post; c NHM; b George Bernard/SPL/photolibrary.com 87tl British Library, London, UK/BAL; br Corel Corp. 88tr The Kobal Collection; cr MEPL; bl The Everett Collection; 89 The Kobal Collection 90tr Xinhua-China/GAMMA/ Picture Media; c OLM; b Bruce Seylem/Museum of the Rockies 91t Paul C. Sereno/University of Chicago; br Robert P. Stahmer 92cl Philippe Plailly/SPL/ photolibrary.com; bc Humboldt-Universität zu Berlin Museum für Naturkunde 93 NHM 94tl Tom & Therisa Stack/TSA; cr Cindy Buxton/BBC Natural History Unit; cr Corel Corp.; bl Dan Burton/BBC Natural History Unit 95b F. Gohier/ ARDEA 96t Jules Cowan/BC; b Kathie Atkinson/AUS 97t F. Gohier/ AUS; b Jan Taylor/BC 98c F. Gohier/ARDEA 98t Stuart Bowey/Ad-Libitum 99t Howard Hughes/Nature Focus; c Stuart Bowey/ Ad-Libitum; b Patrick Cone Photography 100tl Stuart Bowey/ Ad-Libitum; tr F. Gohier/ARDEA; b Field Museum, Chicago 101t GAMMA/Picture Media; b OLM 102t Oliver Strewe/ Wave Productions 103t Ira Block/ NGS 105c NHM; b Ken Lucas/PEP 106tl Dover Royalty Free Images; c Neave Parker/ NHM/BAL; b Mary Evans-Explorer/ AUS 107t Dover Royalty Free Images; b Mary Evans-Explorer/ AUS 108bl Tom & Therisa Stack/TSA; br OLM 109t Oxford University Museum of Natural History, UK/BAL 110c Richard

Nowitz/NGS; bl Stuart Bowey/Ad-Libitum 111t GAMMA/ Picture Media; b Corel Corp. 112–113 Getty Images 124bl Ken Lucas/PEP 129b John Long/Western Australian Museum 131c OLM;bl Phil Chapman/ Seaphot Limited/PEP 135b F. Gohier /AUS 137cl F. Gohier/ARDEA 138b AMNH (Finnin/No. K16693) 144b Humboldt-Universität zu Berlin Museum für Naturkunde 149bl NHM, London 150br Gary Bell/PEP 151c P. Morris/ARDEA; b AAP Image 152c AMNH (Fulda/Neg No. 39122) 154b NHM 155b AAP Image 161b Corbis 162br AMNH (No. K17223) 163c Corbis 164bl RTMP 166bl OLM 167bc OLM 169b OLM 173b Queensland Museum 174b RTMP 177bl Tom and Pat Rich 178c Tom and Pat Rich 179br Tom and Pat Rich 183bl NHM; cr North Wind Picture Archives 185b Ken Lucas/PEP 187b AMNH (Brown/Neg No. 18541) 189b Kevin Schafer/photolibrary.com 190b Queensland Museum 192c NHM 194b Tom & Therisa Stack/TSA 196b RTMP 199t F. Gohier/AUS 202t N.C. Museum of Natural Sciences 203bl OLM 205t Carlos Goldin/SPL/photolibrary.com 209t Jules Cowan/BC 206–207Corbis 210 Chip Clark/ Smithsonian Institution 211 Chip Clark/Smithsonian Institution 212c Gail Mooney/Corbis/ APL 213t Gail Mooney /Corbis/APL; c AMNH (Neg. No. 17837) 214t F. Gohier; c F. Gohier/AUS 215t F. Gohier/AUS; b Jeff Foott/TSA 216t AMNH (Neg No. 19508); c Hammer & Hammer Paleotek; b NCMNS 217t NCMNS; cr Hammer & Hammer Paleotek 218t F. Gohier/ARDEA 219t F. Gohier 220t F. Gohier/ARDEA 221t John Eastcott/YVA Momatiuk/ PEP; c F. Gohier; cr RTMP 222t NHM 223 NHM 224t Musée National d'Histoire Naturelle 225t Corbis/APL; c C. Munoz-Yague/Eurelios 226t Jura Museum, Germany; b Urwelt-Museum Hauff 227t Tom & Therisa Stack/AUS; c Sinclair Stammers/SPL/ photolibrary.com 228 OLM 229 OLM 230 IVPP 231t Xinhua-China/GAMMA/ Picture Media; cr Xinhua-China/GAMMA/ Picture Media 233 Xinhua-China/GAMMA/Picture Media 234 OLM 235 OLM 236t Georg Gerster/WO; b AMNH (Neg. No. 411030) 237t AMNH (No. K17179); c AMNH (No. 322019) 238t F. Gohier/AUS 239t Ed Darack/PEP; c *Ciencia Hoy*; br NHM/Orbis Natural History Museum 240t Peter Steyn/ ARDEA; b K. Van Willingh/South African Museum, Cape Town 241t Sinclair Stammers/SPL/photolibrary.com; b K. Van Willingh/South African Museum, Cape Town 242t S. Wilby & C. Ciantar/AUS; b Queensland Museum 243 Queensland Museum 244t Robert Jones/Nature Focus 245t Robert Jones/ Nature Focus; c Jocelyn Burt/photolibrary.com; b Robert Jones/ Nature Focus 246–247 photolibrary.com

Illustration Credits
Anne Bowman 40tr, 40c, 41b, 170b; **Christer Eriksson** 15b; **Simone End** 38tl, 40b, 65tr, 67t, 67cr, 71bl, 73b, 136c, 188c; **Cecilia Fitzsimons/Wildlife Art Ltd.** 27r, 44b, 46b, 47t, 72tc, 78c, 78b, 117c, 118cl, 120b, 121cl, 122cr, 123b, 125b, 126c, 127c, 128l, 132b, 133c, 137b, 140l, 141c, 142t, 143c,145c, 146b, 147cl, 148b, 153b, 155cr, 156c, 157br, 158b, 159c, 160c, 165b, 168c, 171c, 172c, 175r, 176c, 180c, 181b, 182c, 191cr, 193b, 197b, 199cr, 200b, 201r; **John Francis/Bernard Thornton Artists UK** 83t; **Lee Gibbons/Wildlife Art Ltd.** 54tr; **Mike Gorman** 34tl; **Phil Hood/Wildlife Art Ltd.** 120t, 121t, 126t, 127t, 128l, 132t, 135t, 139t, 146t, 152t, 159t, 164t, 165t, 174t, 196t, 198–199t, 200t, 201t; **Mark Iley/ Wildlife Art Ltd.** 117t, 118t, 122t, 130–131t, 141t, 144t, 148t, 149t, 154t, 156t, 157t, 158t, 160t, 161t, 170t, 172t, 173t, 176t, 185t, 189t, 195t, 197t; **Steve Kirk/Wildlife Art Ltd.** 116t, 119t, 123t, 124t, 129t, 133t, 134t, 136–137t, 140t, 142–143t, 145t, 147t, 150–151t, 153t, 155t, 162–163t, 171t, 180t, 181t, 182–183t, 184t, 186t, 191t, 192t, 194t; **David Kirshner** 46tl, 48tl, 56tl, 62bl, 62br, 63tc, 63c, 63b, 68b, 69bl, 71br, 103b, 104tr, 105tr, 171bl; **Frank Knight** 19r, 20t, 24tl, 57br, 64cr, 64c, 64bl, 65b, 70c, 102br, 119b, 134c, 135cr, 162c, 184b, 186c, 192cr, 195c; **David Mackay** 143b; **James McKinnon** 72b, 77b, 80c, 81b, 82tl, 127br, 139b; **Nicola Oram** 34c, 36c, 38c; **Stuart McVicar** 35b, 36b, 39t; **Colin Newman/Bernard Thornton Artists UK** 41c, 130c, 185cr; **Marilyn Pride** 34b; **Michael Saunders** 36tl, 149cr; **Luis Rey/Wildlife Art Ltd.** 125t, 138t, 166t, 167t, 168t, 169t, 175t, 177t, 178t, 179t, 187t, 188t, 190t, 193t, 202-203, 204-205; **Andrew Robinson/ Garden Studio** 104tl; **Peter Schouten** 42tl, 42b, 43tr, 43b, 49b, 66b, 76tl, 76b, 80tl, 82c, 87c, 152cr; **Peter Scott/Wildlife Art Ltd.**104b; **Ray Sim** 102bl.